A HISTORY OF
INVENTING
—— IN ——
NEW JERSEY

A HISTORY OF
INVENTING
— IN —
NEW JERSEY

*FROM THOMAS EDISON TO THE
ICE CREAM CONE*

LINDA J. BARTH

Charleston · London

THE
History
PRESS

Published by The History Press
Charleston, SC 29403
www.historypress.net

First published 2013

Manufactured in the United States

ISBN 978.1.62619.206.5

Library of Congress CIP data applied for.

This book is dedicated to all who love and appreciate the great state of New Jersey.

CONTENTS

ACKNOWLEDGEMENTS

In the preparation of this book, I explored many sources, but one in particular stands out. I owe a debt of thanks to the New Jersey Inventors Hall of Fame for opening its files and biographies of award-winning inventors and innovators.

In addition, I was privileged to interview C. Harry Knowles of Mount Laurel, who taught me about the interpretation of bar codes and his invention of the hand-held scanner. Vicki Hyman, a superb reporter for the *Star-Ledger*, has provided a wealth of information through her feature articles on many New Jersey inventors and inventions, including Play-Doh, the Holland submarine, the blueberry, Calvin MacCracken, the golf tee and Samuel Sorenson Adams.

I also owe a debt of thanks to Robert Barth, for his love, support, suggestions and photographic skills; Christine Retz, for her wonderful editing skills, pertinent questions and friendship; Sue Poremba, for her attention to detail and lifelong friendship; Virginia Troeger and Don Monroe, for clarifying some technological inventions, especially the transistor; Donald J. Peck, for information about Perth Amboy inventors; Lynne Ranieri of the Millburn-Short Hills Historical Society; Jon Gertner, for allowing me to videotape his presentation on Bell Labs; Richard Veit, for information about Clarence Spicer and the universal joint; Brynne and John Solowinski, for sharing their memories of Jerome Lemelson; Peter Materna, for explaining the patent issues concerning Mr. Lemelson; Kelly Mumber, for finding books for research; Mark MacCracken, for details on his father's inventions; Dawn Firsing-Paris; Carol Simon Levin; Savraj Singh; Edward Eckert, archivist at Bell Labs; and Larry Doolittle, for fact checking the entries on Bell Labs' inventions.

For access to photographs, thanks to the following people and organizations:

Edward Wirth, Thomas Edison National Historical Park, West Orange, New Jersey

Joanne Nestor and Joseph Klett, New Jersey State Archives

U.S. Patent and Trademark Office, Washington, D.C.

Library of Congress, Washington, D.C.

Jason Ispanky and Quentin T. Kelly, Chairman and CEO, WorldWater & Solar Technologies Inc

John Kerry Dyke

James Bintliff, President, Lena Blackburne Baseball Rubbing Mud

William Maloney and the New Jersey Aviation Hall of Fame

Bruce Balistrieri, The Paterson Museum

Ben Tabatchnick and Bud Barry, Plant Engineers, Tabatchnick Fine Foods Inc.

C. Harry Knowles and Al Barrett, Metrologic

Charles Shopsin, *Modern Mechanix* magazine

Philip M. Anderson, Professor of Engineering Physics, Theoretical and Applied Science, Ramapo College of New Jersey

Joseph Earley, Alcatel-Lucent, Integrated Information Solutions, Murray Hill, New Jersey

Rory Britt, Warren Township

Wendy Kennedy, Monroe Township

Janet Crisafi, Bridgewater Township

Jonathan Thorn, Corporate Archivist, Campbell Soup Company

Jon Holcombe, www.sledhill.com

Ben Rose, Greater Wildwoods Tourism Improvement and Development Authority

John McCarthy, POWER HAWK Technologies Inc

Melanie Bump, Historic Speedwell

Whitesbog Preservation Trust

Doris Oliver, Library Special Collections, Stevens Institute of Technology

Bart J. Zoltan

Margaret Carlsen, Twin Lights Historic Site

INTRODUCTION

Band-Aids. Movies. Color television. Bubble Wrap. Bar codes. The modern submarine. What do all of these things have in common? Give up? They were all invented in the great state of New Jersey!

New Jersey is truly the land of inventions. M&M's, solar panels, transistors, flexible film and Graham crackers are but a few of the useful and unique creations from the minds of Garden State residents. Not to mention the 1,093 patents issued to Thomas Alva Edison.

Since Edison opened the first research and development laboratory in Menlo Park, the Garden State has been known as the Innovation State. As Alex Magoun, former director of the David Sarnoff Library, put it, "The state's twentieth-century history is filled with the technologies we take for granted, from electronic television and antibiotics to the transistor and liquid crystal displays."

New Jersey inventors and innovators have changed the lives of people around the world. From the phonograph to the electric guitar, from the telegraph to Telstar and from baseball to college football, hundreds of products and ideas got their start in New Jersey.

What makes New Jersey the state where ideas grow? Is it because we've been home to so many communications and pharmaceutical companies, including Bell Labs, Sarnoff, Johnson & Johnson and Edison's invention factory? Is it our proximity to Philadelphia and New York? Or does the fact that we're the most densely populated state mean that bright people are also densely packed in the Garden State?

Whatever the reason, New Jersey has produced hundreds of thousands of new ideas, including innovations in the fields of transportation (the steam locomotive and the steerable balloon), communications (satellites and cell phones), household improvements (air conditioning and the electric knife), entertainment (movies and the phonograph), food (condensed soup and the ice cream cone) and medicine (streptomycin and the artificial knee).

In 1987, Dr. Saul Fenster and Philip Sperber established the nation's first and, to date, only statewide hall of fame for inventors. In 1977, the U.S. Patent and Trademark Office began comparing by state the number of patents issued. New Jersey companies and residents have received 115,000 patents since that date. Only California, Texas and New York have more, and those states have much larger populations.

Open *A History of Inventing in the Garden State* and discover the unique and fascinating inventions and ideas that have shaped and changed the lives of people throughout the world.

THOMAS ALVA EDISON AND THE
INVENTION FACTORY

To say that Thomas Alva Edison was the most prolific inventor in history is not an overstatement. And he performed the vast amount of his work in New Jersey.

Born in Milan, Ohio, in 1847, Edison worked as a telegrapher in various cities for several years, eventually ending up in Boston. There, in 1868, he invented the electrical vote recorder. After leaving Boston, he arrived in New York City, poor and in debt. While looking for work at the Gold Indicator Company, he was on the scene when its tickertape machine broke down. No one but Edison could fix it, and he was given a job as superintendent. He next took a position at Western Union, where he was in charge of all of the company's equipment. His boss at Western Union, impressed with his employee's mechanical talent, bought all of Edison's new inventions and improvements to the company's equipment for $40,000. Edison invested that sum in a factory in Newark, New Jersey, where he produced improved stock tickers and telegraphic equipment. During this period, he continued inventing, patenting new telegraph systems, paraffin paper, the electric pen (forerunner of the mimeograph), the carbon rheostat and the microtasimeter (a device for measuring small temperature changes).

In 1876, Edison decided to move to Christie Street in Menlo Park, a then-rural area of central New Jersey. On Christie Street, he created the "invention factory," a laboratory for organized industrial research—the first research and development facility—with a handpicked team of chemists, physicists, mechanics and mathematicians. Chief among them were two trusted employees who worked with Edison at the Newark factory: John Kreusi, a Swiss-trained

Thomas Alva Edison. *Thomas Edison National Historical Park.*

clockmaker and machine shop foreman, and Charles Batchelor, Edison's chief mechanical assistant from England. In that same year, Alexander Graham Bell invented the first version of the telephone. Edison's former employer, Western Union, commissioned him to improve the phone. As a result, Edison invented the carbon telephone transmitter "button," which made the telephone a success and led to the development of the microphone and the solid-state diode or transistor, which makes many of today's electronic devices possible.

One of the first major inventions at Menlo Park was the phonograph. In November 1877, Edison's team created a machine that allowed a person to speak into a diaphragm attached to a pin that made indentations on a piece of paper wrapped around wood. The first words Edison successfully recorded on the phonograph were, "Mary had a little lamb." By the next year, this invention was known all around the world, earning Edison the title of "The Wizard of Menlo Park." Visitors came by rail (which was right down the street from the laboratory) to see demonstrations of the phonograph. His business was expanding, so he added buildings to the site, including a machine shop, an office and a library. It was at this time that he coined the term "Invention Factory" for the Menlo Park site.

In 1878, Edison began to investigate ways to make the incandescent light bulb burn for a longer period of time. Joseph Wilson Swan, a British scientist, had demonstrated an electric light bulb in December 1878, and his home was the first in the world to be lit by this invention. After improving his original lamp, he did not receive a patent until November 1880. In fact, many men invented bulbs using different types of wires as filaments, but most burned for only a short time and gave little light.

On October 14, 1878, Edison filed his first patent application for "Improvement in Electric Lights," but he continued experimenting with platinum and other metal filaments. By 1879, he had gone back to the use of a carbon filament and conducted the first successful test on October 22 of that year. The bulb burned for 13.5 hours. He continued to tinker with the design and on November 4 filed another patent application, this one for an electric lamp using "a carbon filament or strip coiled and connected…to platina contact wires." The newer bulbs lasted for 40 hours. Edison and his team lit the laboratory and his home with several of the new light bulbs for Christmas. On New Year's Eve, Christie Street became the world's first street to be lit by Edison's incandescent lighting system. Later, Edison and his team discovered that a carbonized bamboo filament could last over 1,200 hours.

But the light bulb was of little use by itself. To make it useful to the public, Edison and his team had to develop the parallel circuit, an improved dynamo, the underground conductor network, the devices for maintaining constant voltage, safety fuses and insulating materials and light sockets with on/off switches. The men of the Invention Factory had to create every one of these elements, test them thoroughly by trial and error and then manufacture them. Edison spent the next several years creating the electric industry, including a power plant. In the summer of 1882, he began building a large generator plant on Pearl Street in New York City. He then supplied all the office buildings and homes on Pearl Street with about four hundred of his bulbs. On September 4, 1882, hundreds of people gathered to witness a never-before-seen spectacle as the generator was turned on and the street was lit with electricity. This success helped Edison to prove his theory that a central generator station could supply power in bulk.

Thomas Edison created some of the world's most important inventions at Menlo Park. While headquartered there, he applied for about four hundred patents on inventions large and small.

In 1886, Edison started building a new facility in West Orange, New Jersey, and the next year, he moved his laboratory from Menlo Park into the new, much larger complex. He worked here for the next forty-four years of his life,

Above: Edison demonstrates the phonograph before the National Academy of Sciences in Washington, D.C. *Thomas Edison National Historical Park.*

Opposite: A replica of the original light bulb produced in the Menlo Park laboratory. *Thomas Edison National Historical Park.*

continuing to improve his earlier inventions and create new ones. During the first four years at the West Orange facility, he received over eighty patents on improvements on the cylinder phonograph. Throughout his years there, Edison and his team achieved unprecedented accomplishments:

1889: First projection of an experimental motion picture.

1894: First commercial showing of motion pictures with the opening of a "peephole" Kinetoscope parlor at 1155 Broadway in New York City.

1896: While experimenting with the X-ray, discovered by Roentgen in 1895, Edison developed the fluoroscope, which he did not patent, choosing instead to leave it in the public domain because of its universal need in medicine and surgery.

1896: Applied for a patent on the first fluorescent electric lamp. This invention sprang directly from his work on the fluoroscope.

1900: Initiated a ten-year period of work resulting in the invention of the Edison nickel-iron-alkaline storage battery and its commercial introduction.

1901: Commenced construction on the Edison Cement Plant at New Village, New Jersey, and started quarrying operations at nearby Oxford.

1902: Worked on improving the Edison copper oxide primary battery.

1907: Developed the universal electric motor for operating dictating machines on either alternating or direct current.

1910: Initiated a four-year period of work on improving the disc phonograph.

1913: Introduced the Kinetophone for talking motion pictures.

1914: Received patent on electric safety lanterns to be used by miners.

These lamps have contributed in an important degree to the reduction of deaths in mines.

1914: Invented the Telescribe by combining the telephone and the dictating phonograph.

1915: Established plants for the manufacture of fundamental coal-tar derivatives vital to many industries previously dependent on foreign sources. These coal-tar products were needed later for the production of wartime explosives. His work in this field is recognized as having paved the way for the important development of the coal-tar chemical industry in the United States today.

1915: At the request of Secretary of the Navy Josephus Daniels, Edison became president of the Naval Consulting Board. During the war years, Edison performed a large amount of work connected with national defense, particularly with reference to special experiments on over forty major war problems for the United States government.

1928: Presented with the Congressional Medal of Honor by Secretary of the Treasury Andrew W. Mellon.

1929: On October 21, while in the presence of President Hoover, Henry Ford and other world leaders, Edison commemorated the fiftieth anniversary of the incandescent lamp by reenacting its creation.

The great inventor, who over the course of his life received 1,093 patents, died on October 18, 1931, at the age of eighty-four at his home in Llewellyn Park in West Orange. Arthur Palmer, an employee of Edison Industries, composed a eulogy, part of which follows:

> *He led no armies into battle, he conquered no countries, and he enslaved no peoples....Nonetheless, he exerted a degree of power the magnitude of which no warrior ever dreamed. His name still commands a respect as sweeping in scope and as worldwide as that of any other mortal—devotion rooted deep in human gratitude and untainted by the bias that is often associated with race, color, politics, and religion. Of this man, this super-being who defies classification, what more can be said, what greater tribute can be paid than this: "He is humanity's friend."*

BELL LABORATORIES

Just as Edison's Invention Factory amazed the nation in the late nineteenth century, Bell Laboratories excited the twentieth-century world with its spectrum of ideas, innovations and inventions.

In 1925, the executives of American Telephone & Telegraph (AT&T) created the Bell Telephone Laboratories, an industrial organization in which theoreticians, experimentalists, material scientists, metallurgists and engineers worked together. Because AT&T was one of the largest companies in the United States at that time, controlling 90 percent of all telephone business, it had a steady stream of money to hire talented scientists who brought their abilities to Bell Labs. Over decades, their innovations were the building blocks for every kind of electronic device we use today: transistors, silicon solar cells, communication satellites, digital communications, optical fiber systems and the laser.

Innovation and invention are often responses to a need. Early telephone equipment could send signals all the way across the country using the then-current technology of vacuum tubes. But Bell Labs scientists discovered that the vacuum tubes were inefficient, as they were too costly and used too much electricity. A practical, solid-state replacement was needed. The answer, after years of experimentation, was the transistor, which amplifies the electric current.

Once again, in the 1950s, a problem existed. Across the country, repeater stations sent the telephone signals on to the next station. But it was difficult to power these stations, especially in remote locations. This led to the creation of solar cells, which used the energy of the sun to help power the stations.

Inventions and research from Bell Labs have led to the development of the cell phone, the transistor, the helium-neon laser, fax machines, solar cells, digital cameras, communications satellites and motion pictures with sound. The basic research achieved at the labs proves that just a small number of people working on a problem can change the world.

Bell Laboratories was inducted into the New Jersey Inventors Hall of Fame in 1989.

JEROME LEMELSON:
METUCHEN INVENTOR

Jerome Lemelson (1923–1997) was a man of ideas. A longtime resident of Metuchen, New Jersey, he received over six hundred U.S. patents and was one of the most prolific inventors in the United States.

As a boy, Lemelson created a lighted tongue depressor for his physician father. With an early interest in toys, he developed a variation on the beanie, adding a propeller on top. Toy companies, however, were not interested in his ideas because they already had similar products on the market, and this led to some of his earliest patent-infringement lawsuits. An example is a suit against Kellogg cereals. Lemelson submitted to the cereal company an idea for printing a children's mask on the box that could be cut out and worn. Kellogg dismissed the idea, as it had used cut-out masks in the past. Lemelson then obtained a patent for his particular mask and later sued Kellogg when he saw a printed mask on a box of Corn Flakes. Another early invention was a motorized toy car on a flexible track, similar to Mattel's Hot Wheels.

As the computer age was beginning in the 1950s, Lemelson became intrigued by that technology. His Machine Vision Technology, developed between 1954 and 1956, is a combination of computers, robotics and electro-optics that allows assembly-line robots to build cars and other products. Years later, when automakers began using his system without paying licensing fees, Lemelson took them to court. He won the case against a group of Japanese auto manufacturers, who then had to pay him $100 million for using automated systems that were based on the "Machine Vision device" he had invented almost forty years before. Several European and

American automakers soon paid him for the use of his invention, giving Lemelson more than $500 million.

To ease the work of handling many documents, Lemelson invented and patented a video filing system that stored documents on videotape. This led to his creation of an audiocassette drive system, which he licensed to Sony for its Walkman cassette player. In the latter part of his life, Lemelson invented various medical devices, including some that he created while he himself was being treated for cancer. At one time or another, his patents have covered Velcro darts, crying dolls, aspects of semiconductors and the bar-code scanner.

Typically, Lemelson did not conduct much laboratory or manufacturing work, nor did he work with a staff or a university. He was an independent inventor, and throughout his life, he advocated for the rights of independent inventors. He received royalties and licensing fees from many U.S. and foreign companies that used his patents to manufacture their products, which included the videocassette recorder (VCR), the fax machine and the camcorder. Over the course of his life, he received over $1 billion in licensing fees. Some of this money was used to continue his court fights over use of his patents. Much of the money, however, was donated to charitable causes that promote invention. At the Massachusetts Institute of Technology (MIT), Lemelson funded the annual Lemelson-MIT Award of $500,000 for invention and innovation. He also donated to the Smithsonian Institution's National Museum of American History, establishing the Jerome and Dorothy Lemelson Center for the Study of Invention and Innovation to encourage creativity in young people.

Today, the Lemelson debate goes on. Adam Goldman and John Hood summarized the controversial inventor in two separate articles. On the August 20, 2005 broadcast of *ABC News*, Adam Goldman said, "To his many detractors, Lemelson's patents were, in fact, worthless. Lemelson, they say, was one of the great frauds of the twentieth century." But in John Hood's article, "How Business Delivers the Goods," in the July-August 1996 issue of *Policy Review*, he noted, "Jerome Lemelson [was] a great philanthropist, [but] the value of his charitable work could not possibly match the value of his contributions to American society as an innovator and entrepreneur."

WHAT IS A PATENT?

A patent is a legal right granted by the government that confers to inventors the exclusive rights to manufacture, use or sell an invention for a certain number of years. During the lifetime of a patent, the patent owner can prevent anyone else from making, using or selling the invention. After a patent has expired, the invention can be used freely by anyone. The public benefits both from the knowledge that is available in expired patents and from the encouragement of new ideas that result from the incentive of patent protection during the lifetime of a patent.

Patents cover inventions in almost any imaginable field of technology. At the present time, some of the most active areas of patenting are electronics and computer technology and medical and pharmaceutical technology. Applications for patents are examined by the U.S. Patent Office to confirm, among other requirements, that the invention is new and is not an obvious variation of existing inventions. In the United States, hundreds of thousands of patent applications are filed every year, and the number of patents issued since the start of the patent system totals more than 8 million. Some inventions are made by lone individuals, while many others are the products of teams of people working together in an organized manner. Sometimes, patents are licensed or sold to other users or owners. Although applying for a patent is an important early step in working on a new idea, it is also important to understand that turning an idea into a commercially successful product requires much more than just obtaining a patent, and only a small fraction of patents actually make money for their inventors.

THE NEW JERSEY INVENTORS HALL OF FAME

The New Jersey Inventors Hall of Fame (NJIHoF) honors inventors, organizations and others who have contributed to innovation in the Invention State. Led by a board of trustees and committees, the NJIHoF promotes the role of invention in the state's development and the role of inventors in improving society and changing our lives. Now in its third decade, the NJIHoF operated from 1987 to 2002 at the New Jersey Institute of Technology, from 2003 to 2007 with support from the Research & Development Council of New Jersey and thereafter under the aegis of the Stevens Institute of Technology Office of Academic Entrepreneurship. As of January 2012, the NJIHoF has honored 357 men and women for inventive achievements that have significant positive impacts on society.

New Jersey has played a key role in the modern practice of invention, thanks to its unusually rich mix of corporate laboratories, scientists, engineers and inventors. Once home to the great Thomas Edison, who industrialized the process of invention at his factories in Menlo Park (now part of Edison Township) and West Orange, as well as Bell Telephone and RCA Laboratories, the state continues to be a powerhouse of creativity and innovation. It ranks fourth nationally in the total number of United States patents issued while ranking only eleventh in population.

Because New Jersey is the only state with its own inventors hall of fame, the U.S. Patent Office and the National Inventors Hall of Fame in Akron, Ohio, have recognized the NJIHoF for its outstanding organization and commitment to honoring those who seek to improve our lives through technology. In addition to inducting members, the NJIHoF grants Inventor of the Year Awards, Innovator Awards, Special Awards, Corporate Awards and others as deemed appropriate by the board of trustees. For example, a new award category—Graduate Student Awards—was instituted in 2005 to promote the invention process and to stimulate inquiry-based and transdisciplinary learning.

To qualify for induction into the New Jersey Inventors Hall of Fame, an inventor must have lived in the state during the period of his or her inventive project or have worked for a company in the state that sponsored the work. Candidates are ranked on several measures, the most important of which is how well the invention or its patent was commercialized or used and the significance of its impact on society. An Innovator Award and a Graduate Student Award do not require a patent, whereas patents are a prime requirement for other categories. Nominations for the Graduate Student Award category must be submitted by appropriate college professors. Forms may be submitted at any time and held for the next review period. It is customary for New Jersey's governor to recognize the award winners at the organization's annual awards banquet.

For information about the New Jersey Inventors Hall of Fame, please visit www.njinvent.org.

CHAPTER 1

THAT'S ENTERTAINMENT!

Flexible Film—Reverend Hannibal Goodwin, 1887

Beginning in the mid-1800s, people began using the recently invented camera to take photographs of friends, buildings and landscapes. At that time, however, the photographs had to be imprinted on a glass plate using chemicals, requiring careful handling. And if the subject was moving, it was impossible to get a decent shot.

Reverend Hannibal Goodwin, a Newark minister, had been using glass slides to show Bible scenes to his Sunday school students, but the glass slides kept breaking. The good minister wondered what he could do to solve this problem. In 1887, working with celluloid (see Chapter 4) in the attic of Plume House, the rectory of the House of Prayer Episcopal Church, Reverend Goodwin invented flexible film—the same kind of film that would go on to be used in movie cameras for decades. The Reverend Goodwin had sensitized a piece of celluloid to hold photographic images. His 1887 discovery of flexible nitrocellulose film ignited the photographic revolution. Flexible film, which could be produced and stored in rolls, made photography accessible to amateurs and led to the mass production of cameras, film and related equipment.

Unfortunately, the good pastor was never able to enjoy the profits from his invention. George Eastman, founder of the Eastman Kodak Company, went to court to argue that he had already invented the flexible film. A judge

Flexible film, similar to this example, was invented by the Reverend Hannibal Goodwin. *Courtesy of Robert H. Barth.*

eventually ruled that the Reverend Goodwin was the true inventor of the film, but sadly, he had died before the judge's decision was made. His wife received the cash settlement from the court.

The Reverend Hannibal Goodwin was inducted into the New Jersey Inventors Hall of Fame in 1990.

KINETOGRAPH AND KINETOSCOPE— THOMAS EDISON, 1889–91

With the invention of flexible film, inventors in Great Britain, France and the United States began to develop machines to record and project moving images. In New Jersey, Thomas Edison's company created the kinetograph (a motion picture camera) and the kinetoscope (a motion picture viewer). The kinetograph camera used stop-and-go film

This gentleman has paid a few cents to peer into the kinetoscope and view a short movie. *Thomas Edison National Historical Park.*

movement to photograph movies for experiments in the laboratory and eventually for commercial kinetoscope presentations. The kinetoscope looked like a cabinet with a peephole on the top. Inside, film moved on spools. Looking through the hole, people watched black-and-white movies that lasted for about ninety seconds.

On May 20, 1891, Edison's first kinetoscope prototype was demonstrated to a convention of the National Federation of Women's Clubs, which had been invited by Mrs. Edison to visit the laboratory. In June 1892, Edison announced his plan to show his kinetoscope at the World's Columbian Exhibition in Chicago the following year. This was the first commercial motion-picture machine. In some cities, Edison opened stores called "parlors," in which many kinetoscopes were set up for viewing. People paid a few cents to view these early movies. Soon afterward, Edison and others improved the kinetoscope so that the movies could be enlarged and projected onto a screen. In France, the Lumière brothers held the first public viewing of a projected motion picture on December 28, 1895. Thomas Edison produced the first such show in the United States a few months later, on April 23, 1896.

On August 26, 1910, Edison demonstrated for reporters an improved version of the kinetograph known as the kinetophone, a device for showing a movie with synchronized sound. Generally, however, from the 1890s until the late 1920s, movies were silent, except for music that was played in the theater during the show. The "talkies" changed movies forever—now people could hear the actors speaking! By the 1930s, Hollywood movies entertained millions of Americans in thousands of theaters across the country. People paid a nickel or a dime to go to the movies, where they saw cartoons, a newsreel and, sometimes, two feature films.

Thomas A. Edison was inducted into the New Jersey Inventors Hall of Fame in 1989.

MOVIES WITH SOUND—BELL LABS, 1926

In 1926, Bell Labs created the first commercially successful system for recording and reproducing motion pictures with sound. This version produced higher-quality films than those made by other companies.

TALKING MOTION PICTURES—LEE DE FOREST, 1920S

After inventing the triode audion tube (see Chapter 2), Lee de Forest next turned his attention to the development of talking pictures. By the early 1920s, he succeeded in devising an electrical-optical method of recording sound waves on film so that they could be rebroadcast in synchronization with pictures. On April 12, 1923, de Forest presented the first commercial talking picture at the Rivoli Theater in New York City. Although most major film studios ignored the invention at first, by 1926, Warner Brothers, Fox and other film companies began to use it, and a new era was launched.

Born in Council Bluffs, Iowa, Lee de Forest displayed an inventive nature early on, making an improved typebar for his typewriter, an improved compass joint and other inventions. A graduate of Yale's Sheffield Scientific School, de Forest specialized in theoretical mathematics, physics and electricity. De Forest died in 1961 in the town that his invention helped to turn into the film capital of the world, Hollywood, California.

MOVIE SOUNDTRACK—GLENN LESLIE DIMMICK, 1931

Among his ninety-four patents, Glenn Dimmick's most significant inventions came in the areas of sound motion-picture recording, sound-powered telephones, optical lens coatings and dichroic reflectors for color television.

The first attempts at "talking movies" used a phonograph record for the sound. But it was difficult to synchronize the film with the record. A system was needed that could effectively and practically put the soundtrack on the film at the same time the video image was captured. Dimmick's first invention in this field was a galvanometer adapted for recording sound on film. The galvanometer used electrical current from the studio microphones to "wiggle" a small mirror that produced a modulated light beam to the edge of the film.

Working for RCA in Camden, New Jersey, Dimmick invented numerous refinements, including noise reduction on the sound-recording system, making it more effective. The RCA sound-recording system with his galvanometer was still being used by motion-picture studios in 1963 when Dimmick retired, thirty-two years after his initial invention.

Dimmick invented the sound-powered telephone for U.S. Navy ships during World War II. The system, manufactured by RCA, used the acoustic energy from the speaker's voice to generate enough electrical current to power the far-end receiver without a battery.

Glenn Dimmick was inducted into the New Jersey Inventors Hall of Fame in 1995.

Drive-in Movie Theaters— Richard Hollingshead, 1933

Not everyone enjoyed going to the movie theater. Richard Hollingshead's mother was a large woman, too large to fit comfortably in a regular theater seat. It is said that she casually mentioned to her son that it would be wonderful if she could watch movies from the comfort of her car. So in 1928, this New Jersey resident had an idea. What if he had a giant screen installed outside and let people pay to park their cars in front of it? They could watch the movie from the comfort of their own vehicles. And so Richard M. Hollingshead created the world's first drive-in movie theater in Camden, New Jersey.

Hollingshead began experimenting with his idea by hanging a sheet for a screen in the backyard of his home at 212 Thomas Avenue in Riverton. In his driveway, he mounted a 1928 Kodak projector on the hood of his car and projected movies onto the sheet, which he had nailed to trees. He placed a radio behind the screen for sound. To improve his idea, Hollingshead tested the sound with the car windows at different intervals. He tested the theater under different weather conditions, using his lawn sprinkler to produce a "rainstorm." And he liked what he saw and heard.

However, one major problem had to be solved. If cars were parked behind each other, the cars at the rear would not be able to see the whole picture because they would be blocked by the cars in front. This did not stop Hollingshead, who lined up cars in his driveway, spacing them at various distances, and placed blocks under their front wheels until he was able to find the correct spacing and angles. He built ramps to raise the cars' front tires.

In May 1933, after years of experimenting, Richard Hollingshead received patent #1,909,537 for the first-ever drive-in movie theater. Hollingshead formed Park-In Theatres Inc. to show family-oriented movies and spent

$30,000 to set up the world's first drive-in theater on Crescent Boulevard in Camden, New Jersey. It opened to the public on June 6, 1933 with the slogan, "The whole family is welcome, regardless of how noisy the children are." The price of admission was twenty-five cents per car and an additional twenty-five cents per person.

Hollingshead used a sixteen-millimeter film projector to show movies on the white wall of his automotive parts machine shop on what is now Admiral Wilson Boulevard, near the border of Camden and Pennsauken. He called his invention an automobile movie theater. The term "drive-in" did not become popular until the 1950s, when there were around four thousand such theaters in the country. RCA Victor provided three six-foot-by-six-foot speakers to go with the forty-foot by fifty-foot screen. The first movie shown was *Wife Beware*. Hollingshead sold the theater in 1935 and opened another one.

Park-In Theatres licensed the drive-in idea to Loews Drive-In Theatres Inc. but in 1937 had trouble collecting royalties. Eventually, after Loews was taken to court, Hollingshead's patent was ruled invalid in 1950.

DRIVE-IN/FLY-IN THEATER—ED BROWN, 1948

New Jersey also hosted the first combination drive-in/fly-in theater. On June 3, 1948, Ed Brown's Drive-In and Fly-In opened in Asbury Park. Mr. Brown's theater had room for five hundred cars and twenty-five small planes. The drive-in was located next to an airfield so that planes could taxi to the last row of the theater. When the movies were over, Ed Brown provided a tow to take the planes back to the airfield.

THE ADVENT OF TELEVISION

While scientists had figured out how to project pictures in a theater, they were still searching for a way to project pictures and sound across great distances—all the way from a studio to your home. Although television was not invented in New Jersey, it is important to note its beginnings, as later work in the Garden State led to the creation of color television.

John Logie Baird (1888–1946), a Scottish engineer, invented the world's first working television system in Hastings, England, in 1923. In 1939, he showed color television using a cathode ray tube (CRT) and a revolving disc fitted with color filters, a method used by CBS and RCA in the United States. In 1941, Baird patented and demonstrated a system of three-dimensional television. On August 16, 1944, he gave the world's first demonstration of a fully electronic color television display.

Philo T. Farnsworth (1906–1971) loved to experiment with electricity. At the age of twelve, he built the first electric washing machine his family had ever owned. Although they had hoped he would become a concert violinist, his parents accepted that electricity was his chosen field. At Brigham Young University in Utah, Farnsworth researched television picture transmission. (He had already developed ideas for television while in high school.) In 1926, he co-founded Crocker Research Laboratories, later renamed Farnsworth Television Inc.

In 1927, Farnsworth became the first inventor to transmit a television image composed of sixty horizontal lines. The image he transmitted was a picture of a dollar sign. He developed the dissector tube, the basis of all current electronic televisions, and filed for his first television patent (#1,773,980) in 1927. Although he won an early patent for his image dissection tube, he lost later patent battles to RCA. Farnsworth went on to invent over 165 different devices, including equipment for converting an optical image into an electrical signal, the amplifier, the cathode ray, vacuum tubes, electrical scanners, electron multipliers and photoelectric materials. Despite all of his success, Farnsworth shared with his family his feelings about television: "There's nothing on it worthwhile, and we're not going to watch it in this household. I don't want it in your intellectual diet."

In 1923, Russian Vladimir Zworykin invented the iconoscope, a tube used in the first cameras to transmit television. Although it was later replaced, the iconoscope was an essential part of early television cameras. Six years later, Zworykin invented the cathode ray tube, called the kinescope. Zworykin was also one of the first people to demonstrate a television system with all the features of modern picture tubes.

Zworykin studied electrical engineering at the Imperial Institute of Technology in Russia. He was tutored by Boris Rosing, a professor who introduced his student to his experiments of transmitting pictures by wire. Together, they experimented with a prototype cathode ray tube that had been developed in Germany by Karl Ferdinand Braun. Rosing and Zworykin exhibited a television system in 1910 using a mechanical scanner in the transmitter and the electronic Braun tube in the receiver. Rosing disappeared during the Bolshevik Revolution of 1917. Zworykin escaped and briefly studied X-ray technology under Paul Langevin in Paris before moving to the United States in 1919 to work at the Westinghouse laboratory in Pittsburgh. On November 18, 1929, at a convention of radio engineers, Zworykin demonstrated a television receiver containing his kinescope.

Like Farnsworth, Zworykin spoke about his feelings toward television: "I hate what they've done to my child...I would never let my own children watch it."

Vladimir Zworykin was inducted into the New Jersey Inventors Hall of Fame in 1989.

CATHODE RAY TUBE AND ALL-ELECTRONIC TELEVISION RECEIVER—ALLEN B. DUMONT, 1932, 1938

In 1932, working at a small laboratory in the basement of his home in Upper Montclair, Allen B. DuMont (1901–1965) invented the "Magic Eye," a cathode ray tube that could be used as a visual tuning aid in radio receivers. He sold the rights to his invention to RCA for $20,000, which he used as capital for expansion. In 1938, at his Allen B. DuMont Laboratories in Passaic, DuMont developed an all-electronic television receiver. He became an influential member of the National Television Systems Committee, which formulated standards that were ultimately adopted. In 1941, DuMont initiated experimental telecasts over W2XWV (later WABD) in New York.

Allen B. DuMont was inducted into the New Jersey Inventors Hall of Fame in 1998.

COLOR TELEVISION— RCA/SARNOFF LABORATORIES, 1950S

After World War II, both the Columbia Broadcasting System (CBS) and the Radio Corporation of America (RCA) wanted to add color to black-and-white televisions. CBS's first color television design was mechanical, not electronic. Its revolving color wheel caused a flickering image, difficult for the viewer to see. Inventor Vladimir Zworykin had designed not only a kinescope but also a specialized cathode ray that made black-and-white television practical and less expensive.

In 1950, the Federal Communications Commission (FCC) chose the CBS color version as the national standard despite the fact that the system was too "flickery" and did not fit into channels that carried black-and-white television. CBS began color programming, but Americans who already owned black-and-white TVs could not receive the programs. Most did not want to buy a separate set just to see the few color programs that were available.

Finally, RCA's engineers, among them Harold Laws and Alfred Schroeder, came up with a new plan to produce color television. Called a shadow mask, it was a screen full of tiny perforations that combined three electron "guns," one each for green, red and blue. It still needed fine-tuning in people's homes, but the new system worked. In addition, it could be viewed on black-and-white television sets. In 1953, the FCC approved RCA's color television format, and sets that could receive this new format went on sale the following year.

Since 95 percent of the standard for broadcasting and receiving color television signals originated with RCA, we can and should give that company the credit for the invention. RCA also made the sets and produced programs to fill their screens. Within a few years, other companies stopped making color television sets, and few made color programs.

The first commercial television program on color film was an episode of *Dragnet* that aired in December 1953. It was followed by such milestones as a live telecast of the Tournament of Roses parade in January 1954, the first color broadcast of a president (Dwight Eisenhower in June 1955), the first color coverage of the World Series (Dodgers versus Yankees in September 1955) and the first colorcast cartoons (*The Flintstones* and *The Jetsons* in the fall of 1962).

The premier of *Walt Disney's Wonderful World of Color* in September 1961 was a turning point, persuading consumers to go out and purchase color televisions. But it was not until 1966 that NBC became the first network to show the color we all now take for granted on all of its programs. It had taken twenty-five years for color television to go from its earliest prototypes to mass acceptance.

VIDEODISC—JAMES HILLIER, RCA, 1980S

Toward the end of James Hillier's career at RCA, he helped develop the videodisc, a forerunner of the DVD. This system supported the home videos marketed by the company from 1981 to 1986. The project was abandoned as videotapes became more popular.

FLAT-SCREEN TELEVISION TECHNOLOGY—JULIE J. BROWN, LATE 1990S—PRESENT

Julie J. Brown has been a pioneer in developing energy-efficient and organic LED technology. The chief technology officer at Universal Display Corporation in Ewing, Dr. Brown led a team in creating organic LED displays that are being used in both mobile hand-held products and televisions. Today, Samsung is the leader in mass production of this technology, which it incorporates into its popular Samsung Galaxy cell phones. Universal Display Corporation's OLEDs are also being investigated as a next-generation energy-efficient lighting technology.

HDTV—GLENN REITMEIER, 1992

Glenn Reitmeier, a native of Trenton, patented an HDTV compression system in 1992. A researcher and manager at the Sarnoff Corporation, near Princeton, Reitmeier developed this system to make digital television easy to access through the use of a transport layer that meets U.S. standards for both digital high-definition television and the MPEG-2, used for coding movies, videos and music. He holds forty-five patents in digital television technology.

Glen Reitmeier was inducted into the New Jersey Inventors Hall of Fame in 2001.

SOLID-BODY ELECTRIC GUITAR—LES PAUL, 1941

Les Paul became interested in music as a youngster. At the age of eight, he invented a neck-worn harmonica holder that allowed him to play the harmonica while also playing the guitar. As a teenager, he performed at local drive-ins and roadhouses. In order to make himself heard above the noise, he experimented by wiring a phonograph needle to a radio speaker to amplify the sound of his acoustic guitar.

The "Big Band" sound of the 1930s and 1940s spurred the need for amplification, as guitars could not be heard above the brass sections. Early on, Les Paul attached a microphone to his guitar, but he knew that was not the best solution. At the time, many musicians were experimenting

Les Paul with the solid-body electric guitar. *Library of Congress.*

with electric guitars. The first ones, invented in the 1930s, were hollow-bodied instruments. A functional solid-body electric guitar was designed and built by Les Paul from an acoustic guitar made by Epiphone. Paul's original instrument, dubbed "The Log," was actually a four-by-four chunk of pinewood with strings and a pick-up arm. It was one of the first solid-body electric guitars.

Les Paul was inducted into the New Jersey Inventors Hall of Fame in 1996.

IMPROVED GUITAR FRET—DR. PHILLIP J. PETILLO

Dr. Phillip Petillo (1945–2010) held patents on everything from hand-held surgical devices to guitar accessories and improvements to battery-powered hand tools to sensor technology used in motion pictures to high security military projects.

One of Dr. Petillo's most high-profile inventions is an improved guitar fret. A fret is a metal bar placed at intervals along the neck of a guitar. The note created depends on which fret the musician presses the string onto when the string is plucked. The conventional fret is rhombus-shaped, which allows for compression from the fret edge to the center and any place in between. Dr. Petillo's fret design is triangular in shape, improving playability and creating a cleaner, smoother sound. Due to a "space-age" metal alloy, Dr. Petillo's frets and strings also last longer. In addition to holding a number of instrument patents, Dr. Petillo has been constructing, repairing and restoring guitars for over thirty years. His list of clients includes Bruce Springsteen and the E Street Band, Paul McCartney, Sting, Michael Jackson, Dave Mason, June Carter, Linda Ronstadt, Tom Petty and the Heartbreakers, Southside Johnny and the Asbury Jukes, Waylon Jennings, the late Johnny Cash and the legendary Elvis Presley.

Dr. Phillip Petillo was inducted into the New Jersey Inventors Hall of Fame in 2004.

Mr. Edison listens to a later version of his phonograph. *Thomas Edison National Historical Park.*

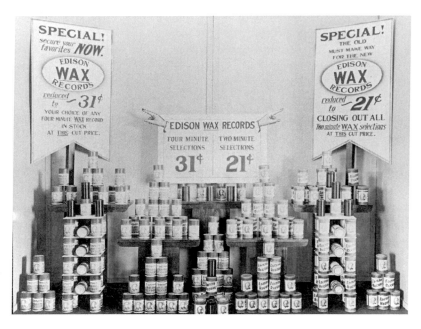

Early Edison recordings were made in cylindrical shapes. *Thomas Edison National Historical Park.*

PHONOGRAPH—THOMAS EDISON, 1877

Before there were CDs and audiocassettes, Thomas Edison invented the phonograph, the first machine that could record sound and play it back. While trying to improve the telegraph, the great inventor noticed that when the paper tape moved through it, the machine made a noise that sounded like spoken words. He designed a phonograph and asked his research team to build it. He then used a cylinder covered with tin foil and a stylus (needle) and spoke into this new machine the first words ever recorded: "Mary had a little lamb." When he played the sound back, he could hear the same words he had recorded.

Edison's first recordings were produced on wax cylinders. The audio recording was engraved on the outside surface. The music or words could then be heard when the cylinder was played on the phonograph. This was the earliest form of recording and reproducing sound. The cylinders, called "records," were most popular from 1888 to 1915. By the 1910s, other manufacturers had switched to disc-shaped records, known as gramophones. The new shape was more successful and popular, but the production of cylinders didn't end until 1929.

MUSIC SYNTHESIZER— HARRY OLSON AND HERBERT BELAR, 1955

A music synthesizer is an electronic console, usually computerized, for creating, modifying and combining tones or reproducing the sounds of musical instruments by controlling voltage patterns. It is operated by means of keyboards, joysticks, sliders or knobs. The synthesizer was developed by RCA employees Herbert Belar and Harry Olson at the David Sarnoff Research Center near Princeton. Olson also designed several of RCA's microphones. Electronic music pioneers, like Milton Babbitt of Princeton University, began composing for the synthesizer. Babbitt worked for RCA to compose for its Mark II Synthesizer at the Columbia-Princeton Electronic Center. In 1961, he wrote *Composition for Synthesizer*.

Sony Walkman—Jerome Lemelson, 1974

In 1974, Metuchen inventor Jerome Lemelson licensed to Sony the drive mechanism for the Sony Walkman, a portable audiocassette player that allowed people to carry music with them and listen through lightweight headphones.

CHAPTER 2

CAN YOU HEAR ME NOW?: COMMUNICATIONS

Since the 1800s, many communication improvements have taken place right here in the Garden State. As early as 1838, Alfred Vail (whose family owned the Speedwell Iron Works) and Samuel F.B. Morse (an artist) invented the telegraph at Speedwell, the Vail estate in Morristown. Today, you can visit the room where this new, high-speed invention was first shown to the public.

Telegraph—Samuel F.B. Morse and Alfred Vail, 1838

At Speedwell Village in Morristown, Samuel F.B. Morse and Alfred Vail invented the telegraph. The village's website prefaces the story with the following quotation:

> *Samuel F.B. Morse is known as the father of the American commercial telegraph, but he would not have succeeded without Alfred Vail as his partner. History glorifies Morse but credits little, or nothing, to Alfred. The reasons are complicated, unfair and more than a little sad.*

Samuel Morse's desire for speedier communication may have been caused by a very personal experience. When his wife died in New Haven,

Connecticut, in February 1825, Morse was in Washington, D.C., painting a portrait of Lafayette. Because communication was so slow, the message took four days to reach him, and the funeral had already been held by the time he returned home. Spurred by the desire to speed messages, Morse began work on the telegraph in 1838 at Speedwell, the estate of Stephen Vail, owner of Speedwell Iron Works.

While all of the books tell us that Samuel F.B. Morse invented the telegraph, that isn't exactly the way it happened. On September 2, 1837, Alfred Vail had happened to see Morse demonstrating his "electro-magnetic telegraph" in New York City. By September 23, Samuel Morse and Alfred Vail had signed an agreement stating that Alfred promised to construct a model of the telegraph by January 1, 1838, "at his own expense" to exhibit before officials in Washington. The Vail family would also pay for other expenses, including the cost of the patent. All patents, however, were to be "taken out in the name and for the exclusive benefit" of Samuel Morse. Although the Vail family shared the profits with Morse, Morse was legally entitled to the major share of the credit.

Before the telegraph was finished, Morse wrote to the U.S. Patent Office, explaining that he wanted copyright protection before his work was completed. His letter specified the following inventions:

> *A system of signs by which numbers and consequently words were signified; a set of type to communicate the signs; a port-rule for regulating the movement of the type; a register which records the signs permanently; a dictionary or vocabulary of words, numbered and adapted to this system of telegraph; and modes of laying the conductors. What I claim as my invention…is a method of recording permanently electrical signs, which by means of metallic wires or other good conductors of electricity convey intelligence between two or more places.*

Vail and Morse did not originally use the Morse code that we know today; Morse proposed a more complicated system. He would give a number to each word in the dictionary. The sender would write out the message in words, look in the dictionary to find the number assigned to each word and then send these numbers by telegraph. At the other end of the line, the receiving machine would make a mark on paper each time the switch was activated. The telegrapher then translated the marks back into numbers and then into words. As their work continued, Alfred, working with his apprentice, William Baxter, began to make significant changes. Morse, working in New York, rarely visited but continued the dictionary work, assigning numbers to over five thousand words.

Finally, on January 6, 1838, the telegraph was ready for a demonstration at Speedwell. The men had covered two miles of wire with cotton as insulation and hung it on nails around the room. Alfred's father came to see him send the message, "A patient waiter is no loser." A few days later, the citizens of Morristown were invited to the first public demonstration as Alfred sent the following message: "Railroad cars just arrived, 345 passengers." The *Jerseyman* newspaper reported, "Time and distance are annihilated,

How did the telegraph work? First, workers from the telegraph company had to install poles and wires between towns and across the nation. At the same time, telegraphers learned the Morse code and practiced tapping out messages. If you wanted to send a message telling Grandma that you were coming by railroad or by wagon train to see her in California, you would walk into the telegraph office in your town and dictate your message to the clerk. The telegrapher then tapped out the message on the transmitter. Each time he pressed the bar, he completed the electric circuit. On the other end, the message would come into another telegraph office, where a worker would translate the dots and dashes into words, write the message and deliver it to Grandma.

and the most distant points of the country are by its means brought into the nearest neighborhood." Late in January, Morse and Vail exhibited the telegraph at New York City University. This time, they used ten miles of copper wire. The inventive Vail had replaced Morse's complicated dictionary system with an alphabet code employing dots and dashes—he had made the telegraph easier to use. Those who witnessed the wondrous machine were impressed. Alfred set up the machine for each demonstration but did not speak in public about his contributions. As stipulated in the aforementioned contract, he let Morse take all the credit.

Congress appropriated $30,000 to build a forty-one-mile test line between the Supreme Court in Washington, D.C., and the railroad station in Baltimore, Maryland. Finally, on May 24, 1843, in ceremonies held in the Supreme Court, Morse sent the message "What hath God wrought," chosen by Annie Ellsworth, the daughter of the patent commissioner. In Baltimore, Alfred waited for the impulses to travel as dots and dashes over the long wires. He had made many improvements to the machine; in fact, it hardly resembled the original telegraph the two had demonstrated back in 1838. Despite the contract he had signed, Alfred attached a note to the base of the machine he used in Baltimore: He stated the he was "the sole and only inventor of this mode of telegraph embossed writing."

In July 1854, Alfred wrote in his journal, "If ever I write the history of the Tel [*sic*], I shall do it honestly, and it will appear what service I have done to the whole concern." Yet he never applied for a patent, nor did he write the story. However, he did write the following:

> *I do not seek renown for myself…I care little for the world's applause. But what I do desire is truth, in relation to the history of the improvements of the Magnetic Telegraph…as may be equivalent to the risk I have run, the interest I have shown, and the improvement I have made in the enterprise.*

Amos Kendall, who knew both Morse and Vail, said, "If justice be done, the name of Alfred Vail will forever stand associated with that of Samuel F.B. Morse in the history and introduction into public use of the electro-magnetic telegraph." Vail's apprentice, William Baxter, said the following:

> *He brought me one day [late in 1837], after working an hour at his drawing, a sketch in which the lever was given an up-and-down movement…and for the first time, we had a mechanism capable of making dots, dashes and spaces. Alfred's brain was at this time working at high pressure and evolving new ideas every day. He saw in these new characters the elements of an alphabet which would transmit language in the form of works [sic] and sentences, and he set about the construction of such an alphabet.*

However, one person wanted the world to have no doubt about the creator of the code. In 1911, in the dark of night, someone—a grandson, it is believed—engraved on Alfred's monument at St. Peter's Church in Morristown the following words:

INVENTOR OF THE TELEGRAPHIC DOT AND DASH ALPHABET

Alfred Vail was inducted into the New Jersey Inventors Hall of Fame in 1989.

LONG-DISTANCE TELEPHONE TECHNOLOGY/CRYSTAL OSCILLATOR—BEATRICE ALICE HICKS, 1942–45

Born in Orange, New Jersey, Beatrice Alice Hicks (1919–1979) became the first female engineer employed by Western Electric, a subsidiary of Bell Telephone, in 1942. (She and many other women had the opportunity to work in a male-dominated industry because so many men had left to fight in World War II.) She worked on long-distance telephone technology and developed a crystal oscillator, which generated radio frequencies. This technology was vital to the aircraft industry. At a time when few women were allowed to enter the engineering field, Beatrice Hicks worked for the advancement and recognition of women engineers.

COMPUTERIZED TELEPHONE SWITCHING SYSTEM—ERNA SCHNEIDER HOOVER, 1950S

Although Erna Schneider Hoover earned a BA with honors in medieval history from Wellesley College, her PhD in the philosophy and foundations of mathematics from Yale University led to her future work. In 1954, she was hired as a researcher at Bell Laboratories, where she created a computerized telephone switching system. The switching system used a computer to monitor incoming calls and then automatically adjusted the calls' acceptance rate. It organized incoming calls to prevent system overload. The principles of Erna Hoover's design are still used today, allowing billions of e-mail messages to be sent each day. A resident of Summit, New Jersey, she was awarded one of the first software patents ever issued (#3,623,007) on November 23, 1971. Bell Labs made her its first female supervisor of a technical department.

FIRST ARTIFICIAL LARYNX—BELL LABS, 1929 AND 1960

In 1929, the scientists at Bell Labs developed a mechanical artificial larynx (voice box). This was a great benefit for people who had lost the ability to speak. After a doctor made an opening in the person's throat, a tube was

This gentleman holds the artificial larynx to his throat, allowing him to speak. *Reprinted with permission of Alcatel-Lucent USA Inc.*

placed against the hole, connecting the mouth and the opening. A metallic reed vibrated inside the tube. Air that was forced up the windpipe went through the tube and across the reed. It was then manipulated in the speaker's mouth to create artificial speech.

This complicated device was replaced in 1960, when Bell Labs developed an electronic version. You may have seen someone holding this device against his or her throat. A vibrator in the device replaces the sounds made by vocal chords, allowing the speaker to form words. The voice sounds unusual, but you can hear and understand the speaker.

First Transmission of Wireless Radio Messages— Guglielmo Marconi, 1899

In 1895, at the age of twenty-one, Guglielmo Marconi worked in a laboratory at his father's estate in Italy and succeeded in sending wireless signals over a distance of one and a half miles using electromagnetic waves. Four years later, while living in New Jersey and working at the Navesink Light Station at Twin Lights in Highlands, Marconi received the first radio transmission from a ship in New York Harbor. Today, a plaque at the Twin Lights State Park reads, "Messages from the first practical demonstration of wireless telegraph were sent from here on September 30, 1899, by Guglielmo Marconi."

Marconi had been invited to America by Gordon Bennett Jr., the owner of the *New York Herald* newspaper, to use his wireless telegraph to report on the America's Cup sailboat race. An antenna mast was erected on top of the north Twin Lights tower. However, the America's Cup race was postponed by a naval review organized by President Theodore Roosevelt to celebrate the victory of Commodore George Dewey's fleet at the Battle of Manila Bay. Marconi's first wireless transmissions in America reported on the progress of Commodore Dewey's fleet. On October 3, the same wireless apparatus was used to receive reports on the America's Cup race between Shamrock and Columbia, just off the tip of Sandy Hook. Marconi maintained a wireless station at Twin Lights for a short time before deciding that other locations along the United States coastline would provide better reception for his wireless equipment.

The Marconi Wireless Telegraph Company of America, incorporated in New Jersey in 1899, consisted of five land stations and forty marine stations

Guglielmo Marconi in a stereopticon view. *Library of Congress.*

by 1908. Marconi's transatlantic radiotelegraph stations were used in pairs. A station near New Brunswick, along the Delaware and Raritan Canal, transmitted while another at Belmar received the weak signals from across the Atlantic. The New Brunswick Marconi Station at JFK Boulevard and Easton Avenue in Franklin Township is today the site of Marconi Park. It had been an early radio-transmitter facility built in 1913 and operated by the American Marconi Wireless Corporation.

After the partial failure of transatlantic telegraph cables, Marconi's facility was taken over by the U.S. Navy in January 1918 to provide vital transatlantic communications during World War I. The New Brunswick Naval Radio Station was the principal wartime communication link between the United States and Europe, using the call sign NFF. President Woodrow Wilson's Fourteen Points speech was transmitted by NFF in 1918. Ownership of the station, along with Marconi's other U.S. stations, was transferred from the navy to RCA in 1920. The antenna masts in Franklin Township were demolished in 1952 to make room for what is now a small mall, but the buildings on the other side of JFK Boulevard were spared. All but one of the brick buildings were demolished around 2004 to make way for a storage-locker facility. The bricks and tiles were saved for use in any future restoration of the spared building.

The Belmar Marconi receiving station was located at what is now Camp Evans. The original buildings were constructed by the Marconi Wireless Telegraph Company of America as part of Marconi's "wireless girdle" around the earth. It was then known as the Belmar Station, which served as Marconi's receiving station, working in tandem with his New Brunswick transmitting station. An operator in Belmar keyed the New Brunswick transmitter, thirty-two miles to the northwest, through a landline connection. The station was closed in 1924 after receiver functions were transferred to RCA's new Radio Central receiver site on Long Island.

In 1909, Marconi shared the Nobel Prize for Physics with Professor Karl Braun. He died in Rome on July 20, 1937.

Guglielmo Marconi was inducted into the New Jersey Inventors Hall of Fame in 1989.

Triode Audion Tube (Birth of Modern Radio)— Lee de Forest, 1900–10

Lee de Forest (1873–1961) earned over three hundred American and foreign patents in radio, telegraphy and motion pictures, but the best known is his "Triode Audion Tube," which for the first time allowed the transmission and amplification of voice and other sound via electromagnetic or radio waves. With the Audion, modern radio was born.

In the early 1900s, the great requirement for further development of radio was an efficient and delicate detector of electromagnetic radiation. John A. Fleming's invention of the electronic valve provided the clue de Forest needed. The most serious difficulty of the Fleming valve was that it was relatively insensitive to changes in the intensity of electromagnetic radiation. Moreover, the Fleming valve could act only as a rectifier and not an amplifier.

The simple but revolutionary innovation that Lee de Forest incorporated in his audion tube was a third electrode inserted between the cathode and the anode, making the tube much more sensitive to electromagnetic radiation. In addition, it acted as an amplifier. Using his Audion Tube, de Forest was able to experimentally broadcast both speech and music.

The promise of the Audion enabled de Forest to raise capital to form a company and begin the process of transmitting and receiving voice. He toured Europe during the summer of 1908, demonstrating and installing radiotelephones and obtaining contracts. On January 20, 1910, in New York City, he demonstrated his new technology by broadcasting a live performance of *The Girl of the Golden West*, featuring Enrico Caruso. The age of radio had begun.

De Forest's invention also made possible long-distance telephone calls. He discovered that if one triode was connected to the input of another and a chain was formed, the triodes could be used to amplify and repeat weak signals. This made intercontinental telephone calls possible. Realizing the implications of de Forest's discovery, American Telephone and Telegraph Company (AT&T) acquired the rights to the Audion.

Lee de Forest was inducted into the New Jersey Inventors Hall of Fame in 2000.

ELECTRONIC CIRCUITS UNDERLYING MODERN FM RADIO, RADAR AND TELEVISION—EDWIN HOWARD ARMSTRONG, 1912–33

Edwin Armstrong (1890–1954), an electrical engineer, invented three of the basic electronic circuits underlying all modern radio, radar and television. While a junior at Columbia University, Armstrong created his first major invention. In the summer of 1912, he devised a new regenerative circuit that yielded not only the first radio amplifier but also the key to the continuous-wave transmitter that still lies at the heart of all radio operations.

During World War I, Armstrong was commissioned as an officer in the U.S. Army Signal Corps and sent to Paris. His assignment to detect possible inaudible shortwave enemy communications led to his second major invention. Adapting a seldom-used technique called heterodyning, he designed a complex eight-tube receiver that, in tests from the Eiffel Tower, amplified weak signals to a degree previously unknown. He called this the superheterodyne circuit, and although it detected no secret enemy transmissions, it is today the basic circuit used in most radio and television receivers.

By the late 1920s, Armstrong set out to eliminate the last big problems of radio static by designing an entirely new system in which the carrier-wave frequency would be modulated while its amplitude would be held constant. Undeterred by current opinion—which held that this method was useless for communications—Armstrong brought forth in 1933 a wide-band frequency modulation (FM) system that in field tests gave clear reception through the most violent storms and as a dividend offered the highest fidelity sound yet heard in radio. It took him until 1940 to get a permit for the first FM station, erected along with a 425-foot tower on the Hudson River Palisades in Alpine, New Jersey. It took another two years before the FCC gave him a few frequency allocations.

FM broadcasting began to expand after World War II, but Armstrong again found himself both limited by the FCC, which ordered FM into a new frequency band at limited power, and challenged by a coterie of corporations on the basic rights of his inventions. Ill and facing another long legal battle, Armstrong took his own life in 1954. Ultimately, his widow won $10 million in damages from infringement suits. By the late 1960s, FM was clearly established as the superior system. The International Telecommunications

Union in Geneva posthumously elected Armstrong to the roster of electrical greats including Bell, Marconi and Pupin.

Edwin Armstrong was inducted into the New Jersey Inventors Hall of Fame in 1998.

FORERUNNER OF THE FAX MACHINE— RICHARD H. RANGER, 1924

In 1924, while working as a designer for RCA, Richard Ranger (1889–1962) invented the wireless photoradiogram, or transoceanic radio facsimile, the forerunner of today's fax machines. A photograph of President Calvin Coolidge sent from New York to London in November 1924 became the first photo reproduced by transoceanic radio facsimile. Commercial use of Ranger's product began two years later. Officials of the Marconi Wireless Telegraph Co. in London cooperated with RCA in introducing the Ranger system, which could send a five-by-seven photograph in about twenty minutes.

In 1928, Ranger invented the photo-radioscope, which used jets of hot and cold air playing on a sensitized screen to record enlarged pictures. Three years later, he set up a consulting business specializing in radio, acoustics and general electronic technique. In 1933, he invented the electronic chimes, an automatic device to reproduce the familiar hand-struck chimes used by NBC. By connecting his electrically operated chimes with outdoor loudspeakers, he was later able to create the effect of church bells.

During World War II, Ranger was put in charge of radar and communications for the U.S. Army and spent the years from 1944 to 1946 on technical intelligence missions in Europe. While in Germany, he studied advances in tape recorders, which led to further development of magnetic tape recorders after the war.

The Academy of Motion Picture Arts presented Ranger with an Oscar in 1956 for his development of the tape recorder and synchronization of film and sound. Ranger was a 1911 alumnus of the Massachusetts Institute of Technology.

Richard Ranger was inducted into the New Jersey Inventors Hall of Fame in 1997.

TOUCH-TONE DIALING—BELL LABS, 1964

First introduced by Bell Labs in 1964, touch-tone dialing used push buttons and electronic tones, replacing rotary dials. (A rotary dial was a circular apparatus containing ten holes, one for each digit. The user would stick a finger into the hole for the first digit of the phone number, drag the circle clockwise to a metal stop and then let it return. He or she would then do the same for each of the other numbers.) Touch-tone dialing is quicker,

A variety of touch-tone phones. The first phones had only ten buttons; the star (*) and pound (#) buttons were added later. *Reprinted with permission of Alcatel-Lucent USA Inc.*

and the electronic tones can travel over the whole phone system. Scientists chose people to test the system, and they discovered the best positions for the buttons to lessen human error and maximize dialing speed. Early touch-tone phones had only ten buttons. In 1968, Bell Telephone added the star (*) and pound (#) buttons, anticipating that they would be needed for computer-based services. Touch-tone dialing led the way to voice mail and other applications.

TELSTAR—BELL LABS, 1962

On July 10, 1962, the communications satellite Telstar was launched in Florida. This 170-pound sphere, created at Bell Laboratories, allowed much better communication between North America and Europe for both telephones and television. Telstar was the first active relay communications satellite. It received a signal from Earth and then transmitted it to another spot, crossing the Atlantic Ocean.

The day after the launch saw the first telephone call transmitted by satellite. From Maine, Fred Kappel, the chairman of AT&T, which owned Telstar, called Vice President Lyndon Johnson in Washington, D.C. They spoke for one minute and forty-two seconds, as Kappel said, "Good evening, Mr. Vice President."

Next, the first image was transmitted to France: it was the American flag. Then the first fax was sent via satellite, with a photo of Telstar sent to Europe.

Telstar was about one yard in diameter and covered with solar cells to provide power. Without the solar panels and transistors (also invented by Bell Labs), Telstar would not have been possible.

DISPOSABLE CELL PHONE— RANDICE-LISA ALTSCHUL, 1999

In 1996, when cell phone use was quite new, Randice-Lisa "Randi" Altschul was often frustrated by poor connections. She was tempted to toss her cell phone out of the car but quickly realized that these phones were too expensive

Telstar, created by Bell Labs, was the first active-relay communications satellite. *Reprinted with permission of Alcatel-Lucent USA Inc.*

to lose. This realization gave her an idea: Why not make a disposable cell phone that people could buy, use and then throw away?

After checking to see if anyone else had already invented one, Altschul worked with engineer Lee Volte to develop super-thin circuitry that would go inside the phones. In 1999, she was awarded a series of patents for the world's first prepaid wireless cell phone and for the super-thin technology (STTM) it needed to operate. Known as the Phone-Card-Phone, the phone is less than half a centimeter thick and is made from recycled paper products. The phone comes with sixty minutes of calling time and a hands-free attachment. It can be used only for outgoing calls. The user may add more minutes or simply toss it away when the minutes are gone.

Altschul's company, Dieceland Technologies, of Cliffside Park, has not yet distributed the phones, since other companies began making disposable phones of their own. Despite this, Altschul is known as the first person to have conceived of this novel idea. (You will learn more about Randi Altschul in Chapter 4.)

Direct Distance Dialing (DDD)—Bell Labs, 1951

Nowadays, we are used to being able to pick up the phone and dial anywhere in the world. But this wasn't always the case. In the early days of the telephone, the caller would dial a telephone company operator, who would then say, "Number, please." The caller then responded with words and numbers, for example: "Pennsylvania 6-5000," "Murray Hill 5-6543" or "Randolph 2-3585." Then the operator would connect the caller to the person he or she was calling. Later on, callers could directly dial their phones to reach neighbors in town, but calls to distant places still required the help of an operator.

But on November 10, 1951, the mayor of Englewood, New Jersey, became the first person to directly phone someone across the country. He called the mayor of Alameda, California, using the new three-digit area code. (To fully realize the improvement, consider this fact: in 1915, the first transcontinental telephone call took five operators twenty-three minutes to set up a call from San Francisco to New York.) Today, we are used to dialing the area code followed by the usual seven-digit phone number, but in 1951, this was a brand-new idea. Combined with automated switching systems, this North

American Numbering Plan assigned three-digit codes across the continent, allowing direct distance dialing (DDD). Created by AT&T and Bell Labs, DDD lowered the cost of long-distance calls while also making them faster. In addition, calls became more private, as the operator no longer had to be on the line.

SECOND LIVE TELEVISION TRANSMISSION IN THE UNITED STATES—HERBERT IVES AND COLLEAGUES, BELL LABS, 1927

Television, the wonderful invention that brings you cartoons, movies, comedies, dramas and the news, was first demonstrated by Bell Labs on April 7, 1927. On that day, reporters and others met at the New York City office of Bell Laboratories to see the first demonstration of television. The voice and picture of Secretary of Commerce Herbert Hoover were transmitted over telephone lines from Washington, D.C., to New York. In his remarks that day, the secretary said, "Today, we have, in a sense, the transmission of sight for the first time in the world's history. Human genius has now destroyed the impediment of distance in a new respect, and in a manner hitherto unknown." A second telecast followed that day via radio transmission from Whippany, New Jersey. The telecasts demonstrated television's potential as an adjunct to telephone service and as a medium for entertainment.

MOBILE CELLULAR COMMUNICATION SYSTEM— AMOS JOEL JR., 1971

As a boy, Amos Joel Jr. (1918–2008) was fascinated by electric toy trains (how they switched tracks) and his family's new dial telephone. He wanted to know how they worked. When Joel joined Bell Labs in the 1940s, mobile phones were so large that the main equipment had to be stored in the trunk of the car—but more troublesome than size was the frequency. Wireless calls are sent and received on radio frequencies, but the number of frequencies is limited. In the early days, the FCC gave priority to emergency and police mobile phones, leaving only a few frequencies for use by the public. "The main thing was to make available more networks for setting up mobile phone calls," said Amos Joel.

Joel's assignment at Bell Labs was to find a way to reuse frequencies. Building on the work of another Bell Labs scientist, he developed a switching system that depended more on higher frequencies, where signals travel shorter distances. His 1971 patented invention relied on the use of multiple, low-frequency transmitters spread in a hexagonal grid throughout a city. As the mobile-phone user travels, the call is handed off from one grid to another and one frequency to another.

For his invention, Amos Joel was inducted into the National Inventors Hall of Fame and received the Institute of Electrical and Electronics Engineers (IEEE) Medal of Honor, the Kyoto Prize and the National Medal of Technology.

Amos Joel Jr. was inducted into the New Jersey Inventors Hall of Fame in 1989.

CHAPTER 3

FOOD, GLORIOUS FOOD:
INNOVATIONS IN THE KITCHEN

The Ice Cream Cone—Italo Marchiony, 1903

Meet Italo Marchiony (also spelled Marchioni and Marcioni) from Hoboken. Marchiony came to New Jersey from Italy in 1895 and sold his ice cream and lemon ice from a pushcart. As his customers placed their orders, he filled dishes with his delicious ice cream—unfortunately, customers kept walking away with the dishes! When he had none left, he had to stop selling his frozen treats. He needed a new plan. Why not make a dish that they could eat? So Marchiony created an edible pastry shaped like a cup. The pastry held the ice cream, so he no longer needed dishes. Today, we call this cup an ice cream cone.

At first, Mr. Marchiony baked waffles and shaped the warm pastries into cups. This waffle cup became so popular that he hired other men with pushcarts to sell his ice cream in a cup. But with so many customers, he couldn't make enough handmade cups to keep up with the demand. In order to speed up production, he designed and built a machine to mass-produce the edible dishes. He changed the design of a standard waffle iron so that it baked the batter in the shape of waffle cups. This was tricky, however, because the fragile cups were easily broken.

On December 15, 1903, Italo Marchiony received U.S. Patent No. 746,971 for an ice cream cone–mold machine, which made small pastry

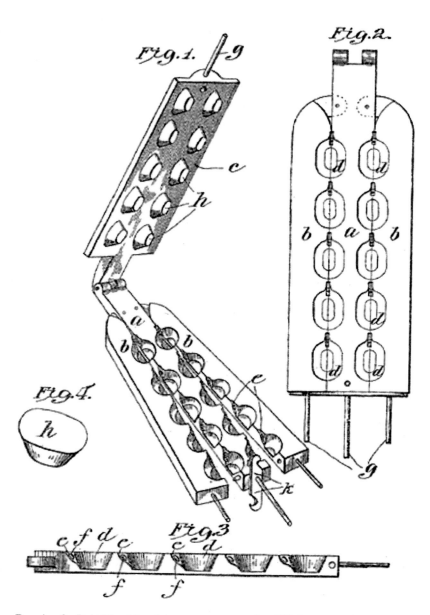

Drawing for Italo Marchiony's ice cream cone maker. *U.S. Patent and Trademark Office, Washington, D.C.*

cups with sloping sides. The patent clearly states that the owner of the patent was "Italo Marchiony of New York," but since he lived in Hoboken and sold his wares in New York, we claim him as a Garden State inventor.

At first, it was difficult to take the fragile cups out of the mold without breaking them. Marchiony solved the problem by dividing the bottom half of the mold to separate it from the baked cups. Instead of one mold for each cup, he arranged two rows of five in each mold to produce ten cups at a time. Marchiony took his confection to the 1904 St. Louis World's Fair (also known as the Louisiana Purchase Exposition). While there, he ran out of his patented cups and asked a waffle maker in the next booth to roll the waffles into the shape of a cone. Because of the success at the exposition, the idea of an edible ice-cream container spread throughout the country. Marchiony's company thrived at 219 Grand Street in Hoboken, turning out ice cream cones and wafers until his plant was destroyed by fire in 1934. He retired from his business in 1938 and died in 1954 at the age of eighty-six.

Now, you may have heard a story about the ice cream cone being invented in St. Louis during the 1904 fair. In this version of the story, the ice cream seller ran out of cups and asked the waffle maker next door to make a waffle that could be rolled into a cone to hold the ice cream. The unnamed ice cream seller was our own Italo Marchiony. Although ice cream cones became popular after they were seen at the World's Fair, remember that Italo Marchiony's patent was granted in December 1903, months before the fair opened.

M&M's—FORREST E. MARS (1904–99) AND BRUCE MURRIE, 1941

Following in the footsteps of his father, candymaker Frank C. Mars, Forrest E. Mars invented M&M's. The elder Mars had created Snickers, Milky Way, Three Musketeers and Mars bars from his own company, Mars Inc., in his home in Tacoma, Washington, in the 1910s. In the 1930s, Frank invited his son to join him in his now-flourishing candy company in Chicago. Forrest wanted to expand the business, but his father did not. Taking a buyout and the rights to sell Mars candy overseas, Forrest moved to Europe.

There he worked for Nestlé and Tobler to learn the candy business and set up a small factory in England. Returning to the United States,

"The milk chocolate melts in your mouth, not in your hand." *Courtesy of Robert H. Barth.*

Forrest started Food Products Manufacturing, establishing the Uncle Ben's line of rice and, later, gourmet pet food. In 1940, he entered into the candy business in the United States, and having already seen a chocolate candy with a hard, sugar shell (possibly during the Spanish Civil War), he concocted his own version of the candy. Conferring with Hershey's Bruce Murrie, Forrest proposed an 80/20 partnership, with Hershey providing the chocolate. This was a good idea, since chocolate would be rationed during the anticipated war, and only Hershey had a contract to provide chocolate for the troops. With this deal, Murrie supplied some capital, and Hershey provided the chocolate, sugar and technology. The partners named the product for the first initials of their last names and added the apostrophe because it belonged to them: M&M's.

When the patent was awarded in 1941, M&M's began production in Newark, making the candy shells in brown, yellow, orange, red, green and violet (later replaced by tan). When World War II began, M&M's were instantly popular with the soldiers, since they could tolerate heat. In fact, the candies were part of the C-rations for the troops.

After the war, Mars bought out Murrie and became the sole owner. Soon after, the company began printing the iconic "m" on each candy so that customers would recognize the real thing. In 1954, Mars introduced peanut candies and launched a new advertising campaign with the now-famous slogan, "The milk chocolate melts in your mouth, not in your hand."

At the time of his death in 1999, Forrest Mars Sr. had an estimated net worth of $4 billion. His sons and daughter are also among the richest people in the country. Now based in Hackettstown, M&M/Mars employs thirty thousand people worldwide with sales of over $20 billion per year.

FIRST CULTIVATED BLUEBERRY—ELIZABETH WHITE, 1912–28

As a young woman, Elizabeth Coleman White (1871–1954) began working on her family's cranberry plantation at Whitesbog, New Jersey, in the Pine Barrens. In 1911, she invited Frederick Coville to come and work at Whitesbog to further his work on blueberry propagation. Until then, New Jersey farmers had felt that cultivation of the fruit was impossible in the Pine Barrens, despite the fact that wild blueberries grew there in abundance.

Blueberries were first cultivated as a crop by Elizabeth White (shown here with Frederick Coville). *Courtesy of Whitesbog Preservation Trust.*

With a team of local people, White and Coville searched for and measured the largest berries and tagged the plants from which they came. Finders answered questions about the fruit regarding plant vigor, resistance to cold and disease, flavor, texture, productivity and time of ripening. White then named the new varieties after their finders, while Coville cross-fertilized the plants to create more new varieties. By 1916, the pair had produced the nation's first commercial crop of blueberries. They were marketed under the name Tru-Blu-Berries.

White also introduced the use of cellophane to package the blueberries, and in 1927, she helped to organize the New Jersey Blueberry Cooperative Association. She also rescued native holly from obscurity, founding the Holly Society of America in 1947.

CRANBURY SORTER/SEPARATOR—D. T. STANIFORD

The problem of sorting cranberries was solved when D.T. Staniford, an Ocean County resident, moved to New Brunswick, New Jersey. He had heard the story of John "Pegleg" Webb, a cranberry farmer near Holmanville. It seems that old Pegleg stored his cranberries on the second floor of his storehouse and, for some reason, sorted them on the first floor. His peg leg made it difficult to carry the boxes of berries downstairs, so he dumped them down the stairway. He quickly noticed that the healthy berries rolled all the way down, while most of the rotted ones stayed on the steps.

With this story in mind, Staniford went to work, using the machines of the shops in New Brunswick. In his sorter, the berries bounced off a pane of glass, which was easier to clean, rather than the wood used in modern machines.

SWEET CORN—LUTHER HILL, 1902

Luther Hill was a farmer in Andover Township, Sussex County, New Jersey, when he developed an heirloom corn seed. The seeds produce five-foot stalks, each containing two six-inch ears of sweet, tender white corn with a sugary flavor.

MASON JARS—JOHN LANDIS MASON, 1858

A Mason jar is a glass jar used in preserving food. These jars had to be sealed so that no air could enter and spoil the food. John Landis Mason (1832–1902), a tinsmith from Vineland, New Jersey, designed a machine that made a zinc cap with threads to seal the jar completely. This invention allowed farm families, as well as those in the city, to preserve produce more easily. Before that, people either had to use inferior containers or preserve their food by pickling, drying or smoking it to save it for winter use. Mason jars come in a variety of sizes, including pint, quart, half-gallon and cup, as well as in wide-mouth and regular-mouth openings.

When he patented the Mason jar on November 30, 1858, Mr. Mason was living in Brooklyn. But since he came from Vineland, he can be included in our list of New Jersey inventors. In addition, according to Adeline Pepper's *The Glass Gaffers of New Jersey*, the first Mason fruit jar was blown at Crowleytown in Atlantic County, New Jersey. Today, the site is a small park on the picturesque Mullica River in Wharton State Forest. The book also notes, "Mason was never a blower but was a Vineland tinsmith who devised the metal seal and rubber gasket that combined to make the best fruit seal up to that date."

FIRST CANNED CONDENSED SOUPS—JOHN DORRANCE, CAMPBELL SOUP COMPANY, 1897

In 1869, a fruit merchant named Joseph Campbell shook hands with an icebox maker named Abraham Anderson and created Anderson & Campbell in Camden, New Jersey. The two men came to have different visions for the company, so in 1877, Campbell bought Anderson's share and expanded the company to sell more than just preserves. The new products included ketchup, salad dressing, mustard and other sauces. One of the biggest sellers was ready-to-serve Beefsteak Tomato Soup.

When Campbell retired in 1894, Arthur Dorrance became the company president. Three years later, Dorrance hired his nephew, chemist John T. Dorrance, at a salary of $7.50 per week. The younger Dorrance even had to bring his own lab equipment. He soon learned that soup, while inexpensive to

A can of Campbell's condensed soup from 1905. *Courtesy of Campbell Soup Company.*

make, was very expensive to ship because it was heavy. He realized that if he could remove the heaviest ingredient, water, the shipping price would drop from $0.30 per can to $0.10! As a result, he created the formula for condensed soup.

Campbell's now divides itself into three divisions: the simple meals division, which consists largely of soups both condensed and ready-to-serve;

the baked snacks division, which consists of Pepperidge Farm; and the health beverage division, which includes V8 juices.

GRAHAM CRACKERS AND BREAD— SYLVESTER GRAHAM, 1829

Dr. Sylvester Graham (1794–1851) became a Presbyterian minister in 1826. He was the pastor of the Bound Brook Presbyterian Church in Bound Brook from 1828 to 1829. Because he wanted Americans to eat healthier food, Dr. Graham invented Graham crackers and bread. Both were made from unsifted flour and had no chemical additives, which he thought were unwholesome. (At this time, bakeries often used additives to make bread look whiter.) Dr. Graham thought that whole-wheat bread was more nutritious and healthy. In Northampton, Massachusetts, where he died, Sylvester's restaurant now occupies the site of Dr. Graham's home.

BOSCO CHOCOLATE SYRUP, 1928

When you need a little energy boost, you might try Bosco, a great New Jersey product! This all-natural chocolate syrup was first marketed in 1928 and is now sold in Europe and the United States. Bosco was created by a doctor from Camden, New Jersey, supposedly to put weight on underweight patients. No one knows exactly who the doctor was, but the chocolate syrup must have been yummy, since the William S. Scull Company bought the recipe and started to make it. In the mid-1950s, Bosco was sold to Corn Products Company, and in 1985, the brand was taken over by Bosco Products Inc., located in Towaco, New Jersey.

Believe it or not, Hollywood has used Bosco for special effects in movies. Alfred Hitchcock used Bosco Chocolate Syrup as fake blood in the shower scene in his masterpiece, *Psycho* (1960). This same technique is believed to have been used in other horror films, including *Night of the Living Dead* (1968). Since the films were black-and-white, the brown syrup looked like blood. Bosco was also part of an episode of the television series *Seinfeld*. Entitled

Bosco has been made in New Jersey
since 1928. *Courtesy of Robert H. Barth.*

"The Secret Code," the episode's plot involved George Costanza's use of
the word "BOSCO" as the secret code for his ATM card because it was
his favorite drink. In another episode, "The Baby Shower," George swears
revenge on an ex-girlfriend who once spilled Bosco on his favorite shirt. Vik
Muniz, a modern artist, has recreated well-known works of art, such as *The
Last Supper* by Leonardo da Vinci, entirely in Bosco Chocolate Syrup.

Roll-A-Grill—Calvin D. MacCracken, 1958

Calvin MacCracken of Englewood, New Jersey, founded the CALMAC Company in 1947. The company produces thermal-energy storage systems to air-condition buildings and equipment for ice-skating rinks. Over the years, other products have included the Roll-A-Grill, a hot dog cooker used all over the world. You've probably seen them at amusement parks, ballparks and other places that cook many hot dogs at a time. To learn more, visit www.calmac.com/aboutus.

Calvin D. MacCracken was inducted into the New Jersey Inventors Hall of Fame in 1989.

Broil-A-Foil—Elihu Tchack, 1950s

Dinner is over, and now you have to clean up. Oh, yuck—look at the burned mess at the bottom of the frying pan! Don't you wish you could just throw it away? In the mid-1950s, Elihu Tchack (pronounced "Chack"), from Livingston, New Jersey, had the same problem. The doctor had ordered his wife to stay in bed during her pregnancy, so Eli was left with the cooking and the cleaning, a job made worse because the family used cast-iron skillets.

As a former toolmaker and the owner of a hardware store in Newark, Tchack (1921–2006) figured he could come up with something better—and he did! He created a disposable aluminum cooking pan and named it Broil-A-Foil. The larger version was nine inches by twelve inches and one inch deep. It was big enough to hold steaks, burgers and chops. And it was cheap—just thirty-nine cents for a box of five pans. But best of all, when dinner was over, you could throw away the pans!

After Tchack received a patent for his Broil-A-Foil, he started a company to make and sell the pans. His Metal Foil Products Company had a factory in Linden, New Jersey, where Eli worked for thirty-five years before retiring in the early 1990s. His pans were sold throughout the United States, but after fourteen years, the patent ran out. This meant that other aluminum makers were now allowed to use his idea and make disposable pans of their own.

A native of Jerusalem, Tchack had come to the United States in 1928. He attended Brooklyn Polytechnic University and took courses in mechanical engineering until he left to join the U.S. Army to fight in World War II. He lived in Livingston for forty-five years before moving to Chatham in 1998.

THE EGG-O-MAT—CAMILLO EPSTEIN, EARLY 1950S

Where do you go to buy your eggs? The supermarket? The local convenience store? Well, for folks in Somerset County in the 1950s to the 1980s, the Egg-O-Mat was the place to go.

Camillo Epstein had a chicken farm at 41 Mountain Boulevard in Warren Township. He and his wife sold eggs from a stand in front of their house. To accommodate customers who needed eggs at night and on the weekend, the farmer built a wooden shed called the Egg-O-Mat. Inside were four refrigerated carousels, containers that turned like a merry-go-round, and Mr. Epstein filled each one with eggs from his chickens. A power line kept the refrigerated eggs cool and the shed lighted at night. A similar invention was created by Heinz Manufacturing Company in Bristol, Connecticut,

but Epstein built his Egg-O-Matt as a convenience for his customers—and to keep them from ringing his doorbell late at night. The contraption dispensed "Eggs of Distinction" twenty-four hours a day. Customers turned a dial to select the egg size they wanted and then inserted coins—a nickel for small, a dime for medium, a quarter

The Egg-O-Mat dispensed eggs to customers in Warren Township. *Courtesy of the Warren Township Historical Society.*

for large and thirty-five cents for jumbo. They would then slide open one of four metal doors to remove a brown cardboard carton containing a dozen farm-fresh eggs. Epstein even added an intercom system to alert him in case someone tried to tamper with the machine.

As development came to Warren Township, Epstein sold part of his farm. The new owner donated the Egg-O-Mat to the Warren Township Historical Society.

SALTWATER TAFFY—DAVID BRADLEY, 1883

In the 1880s and '90s, Atlantic City was one of the most popular beach towns in the country. Alexander Boardman and Jacob Keim had built the nation's first boardwalk in 1870, and people were flocking to the seashore to swim in the Atlantic Ocean.

Although it's uncertain exactly where and when saltwater taffy was first made, it is thought that David Bradley of Atlantic City may have been involved in its creation. The confection first appeared in Atlantic City in the 1880s. Although it is now made by a large number of manufacturers, many people believe that Bradley, a shopkeeper, was one of the original sellers.

A popular story holds that in 1883, a huge storm hit Atlantic City, and the waves flooded the boardwalk. Salt water poured into Bradley's candy store, soaking his supply of taffy. Supposedly, when a youngster asked if he had any taffy left to sell, Bradley jokingly replied that now all he had was "saltwater taffy."

While Bradley may have originated the sweet treat, most people agree that another confectioner, Joseph Fralinger, popularized the candy when he boxed it as a souvenir for tourists to take home.

IMPROVED SIEVE FOR STRAINING HOT FOOD— SARAH E. LASALLE JONES, 1868

In 1868, Sarah Jones (circa 1829–1884) received a patent for an improved sieve for straining hot foods. Jones planned to enlarge her invention to produce jellies, jams and catsups in a factory, but that probably did not happen. Sarah lived in Jersey City with her husband, John D. Jones, until his death in 1879 or 1880.

CORN STAMPER—SYBILLA RIGHTON MASTERS, 1715

In 1715, Sybilla Righton Masters received a British patent for a method of turning corn into cornmeal by stamping (instead of grinding). This was possibly the first patent issued to an American colonist, as there was no patent office in colonial America. Masters planned to build a small mill using horsepower and a larger one using waterpower. It is thought that her husband, Thomas Masters, bought the Governor's Mill to produce cornmeal. Sybilla had married Philadelphia merchant Thomas Masters in the 1690s. The following year, she received a British patent for a method of working straw to make hats and bonnets. The Masterses lived in Burlington Township.

MODERN FRUIT-JUICE INDUSTRY—THOMAS BRAMWELL WELCH, 1869

In 1869, Dr. Thomas Bramwell Welch and his son Charles of Vineland, New Jersey, processed the first bottles of "unfermented wine" for use in the communion service of their church. This event is said to be the start of the modern fruit-juice industry. Today, Welch's is headquartered in Concord, Massachusetts, named for the Concord grape and the home of Ephraim Wales Bull, who developed this American variety. Welch's is part of the National Grape Cooperative Association Inc. and employs over 1,300 people at its American factories.

NUTTY BUTTA—BEN TABATCHNICK, 2009

Ben Tabatchnick of Somerset had read about the thousands of children starving in Africa. Wars and fighting have kept them from finding food. In an area called the Horn of Africa, more than 13 million people need help to survive. Some field hospitals give special milk to children, but not everyone lives near these clinics.

Mr. Tabatchnick, who owns a frozen-soup company in Somerset, thought of a way to help. In 2009, he learned from UNICEF (the United Nations International Children's Emergency Fund) that there was a great need for therapeutic food. So his factory began making a peanut butter paste packed with nutrients. Sold in small foil packets, Tabatchnick's Nutty Butta is easy to distribute to hungry people around the Horn of Africa.

In 2011, the U.S. Agency for International Development bought one thousand metric tons of therapeutic pastes. Half of that amount came from Tabatchnick Fine Foods, while the rest was made by a French company that calls its product Plumpy'nut.

In order not to contaminate the soups and sauces that are made in the New Jersey plant, Mr. Tabatchnick uses a facility in Stone Mountain, Georgia, to make the Nutty Butta.

Nutty Butta is a therapeutic peanut butter paste packed with nutrients and distributed to hungry people around the Horn of Africa. *Tabatchnick Fine Foods.*

Drake's Cakes—Newman Emanuel Drake, 1888

Newman Emanuel Drake (1860–1930) was a philanthropist who lived in Newton, New Jersey. In addition to creating Memory Park and the Newton Theatre, Drake began a specialty baking business after a trip to England, where he learned the recipe for slab cake. At home in 1888, he baked his first pound cake, which he sold by the slice. Drake founded Drake Bakeries in Brooklyn and saw the popularity of his products spread throughout the country. Among its most famous treats are Yodels, Devil Dogs and Ring Dings. In 1980, the company became part of Borden. Eight years later, it was sold to Interstate Bakeries. Today, the aforementioned products are made in Wayne by Hostess.

Microwaveable Pots and Pans—Melvin Levinson

Melvin Levinson of Edison did not invent the microwave oven, but his inventions do relate to that device. Levinson is especially known for two of his many patents: "Two-Stage Process for Cooking/Browning/Crusting Food by Microwave Energy and Infrared Energy" and "Cooking Food in a Food Preparation Kit in a Microwave and in a Thermal Oven." These two patents produced new types of pots and pans that could be used in the microwave as well as in gas and electric ovens. They combined the best features of microwave cooking (speed, low shrinkage, defrosting, proofing and baking) with the best of gas and electrical cooking (browning, crusting, taste and appearance). The new pots and pans came in a kit with a glass cover, a ceramic base, a metal microwave heating grill and an optional metal baking pan. All of the metal components of the kit were dishwasher safe.

Melvin Levinson was inducted into the New Jersey Inventors Hall of Fame in 1998.

AUTOMATIC EGG CANDLING AND GRADING EQUIPMENT— OTTO NIEDERER, 1939

Swiss immigrant Otto Niederer (1890–1978) invented the first practical and commercially successful automatic egg candling and grading equipment. Egg candling shows any shell cracks or blood spots. Niederer established a dairy farm in what is now Washington Crossing State Park in Titusville in 1915, but a series of misfortunes in the late 1930s caused him to lose the farm. Despite this trouble, Niederer and his four sons worked full time to perfect the egg machine. By 1939, they had developed a machine that could weigh and candle 3,600 eggs per hour, a phenomenal speed at the time.

Today, Niederer's basic technology is still in use. Eggs roll along a metal track, first rotating over a candling lamp and then passing over a series of loop scales—metal loops connected to weights of decreasing value. If the egg is heavy enough, it tips the first scale and falls into the jumbo grade. If not, it moves onto the next lighter scale, and so on. Niederer's machine became known as the "Egomatic," but during World War II, it became "Rivomatic," used as equipment to sort aluminum rivets swept from factory floors. After the war, government contracts brought prosperity for the Egomatic. Otto Niederer Sons Inc. developed bigger, more efficient egg machines that not only candled and sorted eggs but also cleaned, counted and packed them. The company folded in 1987 due to large-scale competitors using computer-automated technology.

Otto Niederer was inducted into the New Jersey Inventors Hall of Fame in 1996.

TV AND AIRPLANE DINNERS—WILLIAM L. MAXSON, 1945

A 1921 graduate of the U.S. Naval Academy, William L. Maxson (1889–1947) resigned his commission in 1935. He established the W.L. Maxson Co. in New York City while living in West Orange. He is personally credited with nine inventions. Within his companies, another seventy-two patents were developed. Maxson's best-known invention is an automatic price computer for gasoline pumps. Among his other inventions were a multiplying machine, toy building blocks and various mathematical apparatuses.

During World War II, the "Tuba," as Maxson was known, first developed the idea of heating frozen cooked foods in airplanes ferrying troops overseas. He not only developed the first frozen dinners—later available in commercial flights as "Sky-Plates" and markets as "Strato-Meals"—but also the conventional oven in which to heat them, the Maxson Whirlwind Oven. Today, these meals are known as "TV dinners."

William L. Maxson was inducted into the New Jersey Inventors Hall of Fame in 1992.

Zoku Quick Pop Maker and Zoku Chocolate Station—Yos Kumthampinij, Ken Zorovich and John Earle, 2011

The Zoku Quick Pop Maker freezes pops in as little as seven minutes without electricity. Using their own healthy ingredients, users can quickly make striped pops, yogurt pops or flavored pops. To enjoy Quick Pops at a

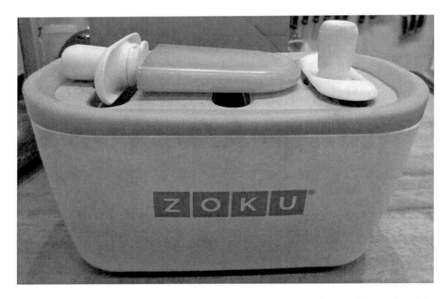

The Zoku Quick Pop Maker freezes pops in as little as seven minutes without electricity. *Courtesy of Robert H. Barth.*

moment's notice, simply store the compact base in the freezer. Once frozen, the Quick Pop Maker can make up to three two-ounce pops before it needs to be refrozen.

Designed to be used with the Quick Pop Maker, the Zoku Chocolate Station allows users to easily and efficiently coat ice pops with hard chocolate shells. Ice pops can be dipped or drizzled with chocolate to create an endless variety of flavors and designs. The unit includes two sprinkle trays that can hold sprinkles or nuts and a specially designed drizzle spoon.

Yos Kumthampinij, Ken Zorovich and John Earle were inducted into the New Jersey Inventors Hall of Fame in 2012.

DRIPSTERS—DAWN FIRSING-PARIS, 2012

Dawn Firsing-Paris said that her young children always ruined their clothes while eating ice cream or ice pops. So a few years ago, after continuing to see stained clothes, she began to brainstorm about creating something to stop the mess. She made some designs in the summer of 2011, but the last

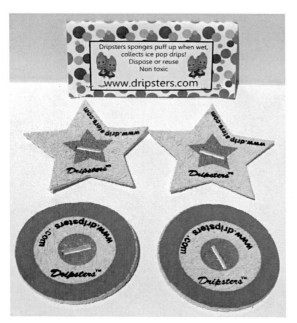

straw was on her family's vacation to Disney World in November of that year, when her son was eating a huge ice cream bar and made a mess of himself. At that moment, she realized she had to get serious about her invention.

Dripsters absorb the drips from ice cream or ice pops. *Courtesy of Dawn Firsing-Paris.*

Once she had designed Dripsters, Dawn had to find a company to manufacture her reusable product. Her research paid off when she came across a Chinese company that could bring her invention to life at an affordable price. Dawn created two Dripsters designs: a round, orange-and-green sponge, and a green, star-shaped one. She knew that bright colors would appeal to children. The stick of the ice pop is inserted through a slot in the sponge, and the sponge absorbs the drips. The Dripster can be washed and reused or discarded.

CHAPTER 4
IT JUST KEEPS GETTING BETTER:
LIFESTYLE INNOVATIONS

Modern plastics were invented mainly in the 1900s, but the earliest kind of plastic—celluloid—has been around for a long time.

CELLULOID—JAMES BROWN AND JOHN WESLEY HYATT, 1870

In the mid-1800s, a New York company, Phelan & Collander, made pool tables and supplies. The pool balls were made from ivory that hunters brought from the tusks of African elephants. Because so many elephants had been killed, there was not enough ivory for all of the balls needed in pool parlors in the 1860s. When the company offered a $10,000 prize to anyone who could invent a substitute for ivory, local printer John Wesley Hyatt took up the challenge. Hyatt rented an apartment in Newark. Although his landlady was not happy about it, he experimented in the kitchen, trying many different mixtures of chemicals. One day, by accident, John combined pyroxylin (a highly flammable substance) and camphor (a gum-like material). Using heat and pressure, John and fellow printer James Brown created a new substance that they called celluloid. Celluloid became the world's first commercially successful plastic. But Hyatt and Brown did not stop with pool balls. They built a five-story factory in Newark and used celluloid to make dental plates, knife handles, piano keys and collars and cuffs for shirts. This last invention made it easier to wash men's shirts. On shirts, the collars and cuffs always got the dirtiest. So men began to attach celluloid (plastic) collars and cuffs to the shirts. These detachable, plastic parts could be easily wiped off with a damp cloth, without washing (and then ironing) the whole shirt. (Remember that doing laundry in those days was a lot more time-consuming than it is today.)

BAKELITE—LEO BAEKELAND, 1907

In 1907, Dr. Leo Baekeland (1863–1944), a Belgian chemist, developed a hard, moldable plastic substance that he named Bakelite. Technically, Bakelite is a thermosetting phenol formaldehyde resin, formed from an elimination reaction of phenol with formaldehyde, usually with a wood flour filler. But we don't have to know all of those details. The important thing to note is that Bakelite had nonconductive and heat-resistant properties and was used extensively in the radio and television industries. It was used to make insulators and casings for telephone wires.

Leo Baekeland invented Bakelite, an early plastic that was used extensively in the radio and television industries. Stoves, knife handles and jewelry were also made of Bakelite. *Courtesy of Somerville Center Antiques and Robert H. Barth.*

In 1910, Baekeland was the president of the General Bakelite Company in Perth Amboy, New Jersey. The company merged with Union Carbide and Carbon Corporation in 1939.

In recent times, Bakelite jewelry has become collectible. The substance was also used to make kitchenware, pipe stems and toys. An early electric guitar, the Electro Spanish, was made of Bakelite and sold by the Rickenbacker guitar company.

PERMACEL TAPE—CYRUS W. BEMMELS, 1949 AND BEYOND

In 1946, Cyrus Woodrow Bemmels (1912–93) conceived the idea of making very strong, thin adhesive tapes by embedding parallel strands of continuing filament yarns in the adhesive. This kind of tape is now commonly referred to as strapping tape. Bemmels's colleagues at Permacel, formerly a division of Johnson & Johnson, in New Brunswick were slow to realize the importance of his discovery, but Bemmels persisted in championing his invention. In 1949, it was placed on the market and was an immediate success.

Bemmels's invention resulted in tape that was five to ten times stronger than previous tapes, without lessening the thickness or flexibility. The tapes replaced steel strapping as well as string and wire. They are good for heavy jobs like bundling conduit or pipe and protect cardboard containers from tampering.

In 1951, Bemmels's first patent was granted for pressure-sensitive adhesives. Over the next twenty-four years, he received eleven more patents for additional tapes and tape-making processes. Bemmels worked on two hundred other tapes at Permacel, especially cellophane and electrical tapes. At one point, his strand-reinforced tapes brought Johnson & Johnson more revenue than any other single invention except the Band-Aid.

Cyrus W. Bemmels was inducted into the New Jersey Inventors Hall of Fame in 1999.

Bubble Wrap—Mark Chavannes and Al Fielding, 1957

In more recent years, inventors used plastic for some very innovative creations. In 1957, Marc Chavannes and Al Fielding were working in a garage in Hawthorne, New Jersey, trying to make plastic that could be applied to wallpaper. As it turned out, they were working on a product that the industry no longer wanted. They quickly realized, however, that their invention could be used for something else. It could be wrapped around items that were going to be shipped to cushion and protect them. At that time, only rough paper products were used for packaging, and that was not good for protecting heavy or delicate items. And so, Bubble Wrap brand protective packaging was born!

Chavannes and Fielding took two sheets of clear plastic film and pumped air between the sheets, making bubble-like pillows. These pillows absorb the bumps and bounces as the package goes from truck to train to plane, and Grandma Alice's teapot gets to Cousin Emily in one piece.

Bubble Wrap is a protective plastic used in shipping. And it's such fun to pop the bubbles! *Courtesy of Robert H. Barth.*

In 1960, Chavannes and Fielding started Sealed Air Corporation to produce their Bubble Wrap brand, and today, Sealed Air is a leading global manufacturer of a wide range of food and protective packaging materials and systems, with annual revenues exceeding $3.8 billion.

Marc Chavannes and Alfred Fielding were inducted into the New Jersey Inventors Hall of Fame in 1993.

The Boardwalk—Alexander Boardman and Jacob Keim, 1870

Sand! It was everywhere on the New Jersey shore. In Atlantic City in particular, the owners of trains, restaurants and hotels wanted to find a way to keep the sand out of their establishments. Alexander Boardman, a railroad conductor, and Jacob Keim, a hotelier, put their heads together and

Alexander Boardman, a railroad conductor, and Jacob Keim, a hotelier, conceived of the idea of constructing a boardwalk as a means of keeping sand out of the railroad cars and hotels. *Library of Congress.*

came up with a solution. Why not build a walkway made of boards? When beachgoers walked across it, the sand on their feet would fall through the cracks and back onto the beach, keeping the hotels, trains and restaurants sand-free. In April 1870, the city council approved an expenditure of $5,000 for the construction of a walkway on the beach. This new boardwalk opened to the public on June 26, 1870. It was eight feet wide, one mile long and stood about one foot above the sand. Twenty-five years later, Boardwalk was made an official street name in Atlantic City. (As a result, it is always capitalized when referring to the street in Atlantic City.)

Before the Great Atlantic Hurricane of 1944, the Boardwalk was about seven miles long and extended from Atlantic City through Ventnor and Margate to Longport. Today, it measures just over four miles. At its widest point, it is sixty feet wide, and it stands twelve feet above sea level. The combined length of the current Atlantic City and Ventnor boardwalks is nearly six miles.

ELECTRONIC ADVANCES MAKE LIFE EASIER

The Bar Code—Norman Joseph Woodland, 1952; George Laurer, 1971

Back in 1948, when he was teaching mechanical engineering at the Drexel Institute of Technology, Norman Joseph Woodland (1921–2012) became interested in the supermarket checkout process. If stores could automatically store and retrieve the price of each item, it would speed customers through checkout lines and simplify running the store. Woodland quit his job to focus on developing ideas for coding products. After trying different ideas, he came up with a system that used elements of two technologies: the Morse code and movie soundtracks. With the help of electrical engineer Bernard Silver, Woodland, an Atlantic City native, tried making lines but then switched the design to concentric circles. In 1949, the two men applied for a patent for bar-code technology. The patent was granted in 1952. While awaiting the patent approval, Woodland and Silver built the first actual bar-code reader. This prototype was as big as a desk, and its five-hundred-watt bulb made the paper smolder, but it worked. Unfortunately, their invention was too big and wasted too much heat to be practicable. Its time had not yet arrived.

The bar code was originally created in a circular shape but later changed to the more effective rectangular shape. *Courtesy of Robert H. Barth.*

Woodland joined IBM to work on his idea. The company offered to buy his patent, but at too low a price. By the 1970s, however, RCA and other companies were experimenting with circular bar codes for the grocery industry. Finally, IBM realized the inventor of this technology still worked for them, and Woodland began to play an important role in creating what would become the Universal Product Code (UPC). In 1971, George Laurer, an engineer at IBM, was asked to design a code that could be printed on food labels and read by scanners being developed for supermarket checkout counters. He was told to model it on the bull's-eye-shaped optical scanning code designed in the 1940s by Woodland. But Laurer knew that the circular shape wouldn't work; it would not print well on a high-speed press. Instead, he developed the familiar pattern of stripes that we know today. After low-powered laser technology became practical, the UPC was adopted by the supermarket industry in 1973.

Norman Joseph Woodland was inducted into the New Jersey Inventors Hall of Fame in 1996.

Helium-Neon (HeNe) Laser—Alan White, 1962

Although the first lasers were invented in the late 1950s, their light was invisible. Bell Labs employees Alan White and J. Dane Rigden developed and patented the helium-neon (HeNe) gas laser, the first visible light laser. It is the red laser dot that we see in supermarket scanners and pointers used in presentations, as well as many other uses.

For his HeNe laser (the acronym comes from **L**ight **A**mplification by **S**timulated **E**mission of **R**adiation), White was able to get a mixture of helium and neon to radiate in the visible portion of the electromagnetic spectrum. To do this, he built a laser cavity that was optimized for creating a red light. With a few other adjustments, White and Rigden made the light more monochromatic, or "more red."

Alan White was inducted into the New Jersey Inventors Hall of Fame in 2000.

Hand-Held Bar-Code Scanner—C. Harry Knowles, 1975

In the 1970s, Metrologic, a company founded by C. Harry Knowles in Blackwood, New Jersey, invented the first triggered hand-held bar-code scanner, the X-Scanner. Metrologic made and sold about forty-eight Verifier 315 scanners, a scanner with a portable head for scanning and checking bar codes in print shops.

In 1953, after earning degrees at Auburn and Vanderbilt, Knowles worked at Bell Laboratories, where the transistor had been invented. He worked on diodes, publishing the first papers on voltage breakdown prediction of diffused junction Zener diodes and the first paper on the theory of Hyperabrupt P-N Junctions, used in many fields.

In 1968, after a decade with Motorola Semiconductors, Knowles started Metrologic Instruments. With no experience in optics or lasers, he developed the techniques for building Helium-Neon (He-Ne) lasers with ideas from Scientific American articles and from library research. Working with Norm Edmund of Edmund Scientific, he developed a mail-order-type He-Ne laser. Then, in November 1969 while working with Les Solomon, tech editor for *Popular Electronics*, Harry introduced a He-Ne laser kit that was, at that time, the magazine's most popular kit project. By 1972, Metrologic was the world's largest producer of He-Ne lasers, specializing in the education and kit markets via direct marketing and through distribution.

Making lasers and other optics for bar-code-scanner manufacturers in the early 1970s, Knowles decided in 1974 that Metrologic should develop its own bar-code scanner. He introduced a programmable high-speed bar-code monitor for scanning milk cartons and plastic packaging. It was tested at Weyerhaeuser in Pennsauken in January 1975, and ultimately about fourteen systems were sold. At the NYC Packaging Show in 1976, Metrologic showed the first triggered hand-held bar-code scanner, the X-Scanner.

A later version of the hand-held bar-code scanner invented by C. Harry Knowles. *Courtesy of C. Harry Knowles.*

Debuting in 2000, Metrologic's MS-9540 Voyager, featuring push-button data transmission, is a single-line handheld laser scanner activated by an infrared sensor. The scanner also acts as an automatic presentation scanner when in its stand. The Voyager became one of the best-selling devices of its kind in the world and remains a top product for its owner.

C. Harry Knowles was inducted into the New Jersey Inventors Hall of Fame in 1995. In 1999, Metrologic was inducted as a corporation.

Security Tags—Philip Anderson, 1987

You've probably seen these white plastic security tags hanging from clothing in stores—but you probably didn't realize that they are part of an electronic surveillance system invented by Philip Anderson of Madison, New Jersey.

Security tags like this one were invented by Philip Anderson as part of an electronic surveillance system for stores and hospitals. *Courtesy of Robert H. Barth.*

This invention helps prevent theft and shoplifting. In 1987, as president of Identitech Corporation, Anderson developed this technology that uses amorphous metal. The electronic article surveillance system consists of two parts: a small strip of amorphous metal attached to an item, and two electromagnetic sensors positioned near an exit. When someone leaves the store and passes the sensors with the security tag still attached to the item, the metal strip within the tag begins vibrating. This movement disturbs the detection system's electromagnetic field and triggers an alarm to warn store personnel of a possible theft. Hospitals use other versions of this security system to protect patients. More than 164,000 of these systems are installed worldwide. Many storeowners say that Anderson's invention is their top choice for security.

A professor of physics at Ramapo College since 1990, Anderson teaches introductory and advanced physics, electronics and invention courses. He has twenty-nine U.S. patents and more than one hundred patents worldwide covering amorphous metals, sensors and medical, automotive and security devices.

Philip Anderson was inducted into the New Jersey Inventors Hall of Fame in 2001.

HOUSEHOLD IMPROVEMENTS

Improved Clothes Dryer—Sarah E. Strickland, 1869

Born in Massachusetts around 1812, Sarah E. Strickland moved to Vineland, New Jersey, in 1865. Four years later, she and her neighbor Darwin Crosby received a patent for an improved clothes dryer that would dry dresses without wrinkling them. In 1870, her farmland, on which she grew fruits and vegetables, was valued at $2,500. Sarah Strickland died in 1872.

Vacuum Cleaner—David T. Kenney, 1907

The *New York Times* called him the "father of the vacuum-cleaner industry." Born in 1866 in Whitehouse, New Jersey, David T. Kenney began applying for patents in the early 1900s. Before the patents were approved for small, portable vacuum cleaners, David created a vacuum system for large buildings. He installed a four-thousand-pound steam engine in the basement of a large building in Pittsburgh. Hoses and pipes were attached to the engine and reached to all floors to suck out the dust. He created the Vacuum Cleaner Company and manufactured the devices on North Avenue in North Plainfield, New Jersey. Soon, other businesses began buying his central cleaning systems for buildings. By this point, Kenney had switched from steam power to electricity.

While awaiting his patent approvals, Kenney had asked the Sisters of Mercy to pray for him. He later donated a gift of mountaintop land in the Watchung Mountains, now the home of Mount Saint Mary Academy. The school began in 1908 with elementary and high school classes. Located at Route 22 West and Terrill Road in Watchung, New Jersey, it is now a girls' secondary school.

Before his interest in vacuums, Kenney had his own plumbing business in Plainfield. He patented a "Flushometer" (used to flush toilets), from which he made a considerable amount of money. But his real interest was in helping housewives save time by making housecleaning more mechanized. It had been proven that suction could remove dust from plush or upholstered furniture and carpets, so Kenney invented a vacuum cleaner on wheels and in his patent application called it an "Apparatus for Removing Dust." This was the first of his nine vacuum-cleaner patents. Finally, in 1907, Kenney received the patent he had applied for in 1901. He later received patents for improvements to the vacuum cleaner. These included better nozzles and hoses and even a see-through chamber that allowed one to see the dust deposited in the water. Although others had patented electrically operated vacuum cleaners, Kenney knew that many towns did not have electricity, so vacuum cleaners had to be operated by hand and had to have nozzles and other devices patented by Kenney. The patents expired in the 1920s, but by then, over one million units had been sold.

Lillian Moller Gilbreth and Frank Gilbreth

Inventor, author, industrial engineer, industrial psychologist and mother of twelve children! What more could one woman do? Lillian Gilbreth of Montclair invented many household items that we still use today. Among them are the foot-pedal trashcan, an electric food mixer and shelves inside refrigerator doors.

While raising four young children, Gilbreth (1878–1972) received her doctorate from Brown University in 1915. Her dissertation, "Psychology of Management," stressed the psychological aspects of industrial management, the importance of human relations in the workplace and the importance of understanding individual differences among workers. In the business world, Gilbreth and her husband, Frank, studied the work habits of manufacturing and office employees to find ways to increase output and make their jobs easier. She was one of the first researchers to understand that stress and lack of sleep have a serious effect on workers. While working for General Electric, Gilbreth interviewed over four thousand women to determine the best height for stoves, sinks and other kitchen items. She was a pioneer in the field of ergonomics, an applied science concerned with designing and arranging things people use so that the people and things interact most efficiently and safely; that is, the science of fitting the workplace to the worker to prevent injuries and illnesses or other hazards. These hazards later became the subjects of a book, *Cheaper by the Dozen*, which described how the Gilbreths applied scientific management to the running of their home. Frank B. Gilbreth (1868–1924) was a pioneer in the field of scientific workforce management, positively affecting engineering, education and personnel procedures.

Lillian Gilbreth was inducted into the New Jersey Inventors Hall of Fame in 2005. Frank Gilbreth was inducted into the New Jersey Inventors Hall of Fame in 2006.

Air Conditioning—Willis Haviland Carrier, 1902

Willis Carrier invented the basics of modern air conditioning when he built his first air treatment device for a printing company. Humidity changes in summer affected the paper's size, and as a result, the colors on the papers did not align properly. This first air treatment device was installed in 1902, and the patent for "An Apparatus for Treating Air" was granted in 1906. In that same year, some U.S. textile mills bought the air-conditioning units to combat the extreme heat in the mills. Other industries soon found uses for "conditioned air." Encouraged by this success, Carrier and six friends scraped together $35,000 and formed the Carrier Engineering Company in 1915 in Newark. The company began manufacturing its own products in 1922, at which point Carrier developed the centrifugal refrigeration machine, one of the most significant achievements in the industry's history. This was the first practical method of air-conditioning large spaces.

Air conditioning as we know it today made its debut in 1924. Many Americans' first exposure to air conditioning was in movie theaters that were trying to keep customers coming during summer months. In the late 1920s, Carrier developed smaller "unit air conditioners" for small- and medium-sized businesses. In 1928, the company developed the "Weathermaker," which regulated year-round air temperature, moisture, circulation and cleaning in homes. After World War II, the housing industry began to expand into the suburbs, and homeowners wanted air conditioning. About 430,000 homes had central air in 1955. Thirty years later, air conditioning was included in 70 percent of all new U.S. homes and nearly 90 percent in the American South.

Willis Carrier was inducted into the New Jersey Inventors Hall of Fame in 1999.

Electric Knife—Jerome Murray, 1964

An electric knife is a hand-held knife with two serrated blades powered by a small motor. When the motor is turned on, the blades move rapidly back and forth, creating a sawing motion. The user can then slice the food with little physical effort. Jerome L. Murray (1920–98) invented this knife and seventy-five other items, including the peristaltic pump (used in heart surgery) and the first windmill used to produce electricity, which he invented at the age of fifteen. Electric knives are now used in complicated medical procedures.

Jerome Murray was inducted into the New Jersey Inventors Hall of Fame in 1991.

Gas Furnace—Alice Parker, 1919

In 1919, Alice Parker of Morristown invented a new and improved gas furnace that provided central heating. Parker, a black inventor, was awarded her patent on March 29, 1919.

Safety Cage Light—Fred Topinka, 1996

When the power went out in the New York City building where he worked, Fred Topinka worried about how construction workers could get to safety. He imagined a hanging lamp enclosed in a plastic cage—it would glow in the dark and lead workers to safety. In 1996, he was granted a patent for the Safety Cage light, which can be used with any color lamp and will continue to glow yellow-green for about twenty minutes after the lights in the building go out. In addition, the Safety Cage has an illuminated horizontal arrow that points the way to the nearest exit.

Fred Topinka was inducted into the New Jersey Inventors Hall of Fame in 2001.

Roller Shades—Stewart Hartshorn, 1885

Stewart Hartshorn of Short Hills created and manufactured an improved roller shade for windows. He perfected the spring mechanism that allows the shade to roll up again. His invention was patented in 1885. Six years later, he applied for and was awarded a patent for attaching window shades to rollers. Hartshorn made further improvements and in 1911 received a patent for a way to attach fabric to the rollers. After making his fortune in spring roller shades, Stewart Hartshorn acquired 1,552 acres to build his ideal village, called Short Hills, the first planned commuter suburb in America. He wanted to create "a harmonious community for people who appreciate nature" and a place "where natural beauty would not be destroyed by real estate developments, and where people of congenial tastes could dwell together." His vision is still much of what Millburn/Short Hills is today.

Laundry Detergent Measuring Container—J. Thomas Jennings, 1980s

One day, while watching his wife pour laundry detergent into the washing machine without measuring the specified amount, J. Thomas Jennings of Short Hills, New Jersey, knew there had to be a better way. Jennings spent many hours over the next three years cutting apart and reconfiguring hundreds of plastic containers trying to design a built-in measuring device. The result of his experimenting was a self-contained measuring chamber within the detergent container itself. This invention, patented as the Tip "N" Measure dispensing container, is available in a variety of styles and sizes. It has a "drain-back" feature that allows excess fluid in the measuring chamber to drain back into the container. A precise amount of fluid is dispensed. Jennings has received other U.S. and foreign patents for other measuring and dispensing devices.

In the early 1980s, Jennings founded Container Manufacturing Inc. to manufacture his invention. The company first began in South Plainfield but in 1985 expanded to Middlesex, where the company continues to manufacture these unique containers twenty-four hours a day, seven days a week. Since then, Jennings has received over a dozen U.S. and international patents. His new measuring device eliminates spillage, waste and misapplication of any concentrated liquid product and minimizes the chance for human contact with hazardous materials such as insecticides and weed control concentrates. In addition to its measuring capabilities, the package is safer because the Tip "N" Measure is not opened until the user is ready to dispense the measured amount of product needed. The containers are 100 percent recyclable and easy to rinse before disposal.

It should be noted that Jennings was no stranger to design and engineering. Between 1956 and 1972, Jennings spent considerable time developing better ways to manufacture wire and cable for our telephone systems. His tension control and wire-winding development became the standard for manufacturing wire. At the time, Jennings owned Eastern Equipment and Controls Corporation in Union, New Jersey, where he built this machinery for well-known companies such as Western Electric, Singer, Okonite, RCA and Phelps-Dodge. Most important was the fact that his developments in the field of wire and cable produced telephone wire that significantly reduced the potential for static when making long distance phone calls. Of course, this was long before fiberoptic and cell systems had entered the market.

J. Thomas Jennings was inducted into the New Jersey Inventors Hall of Fame in 2001.

Rotating Lawn Sprinkler—Eugene Beggs and A.S. Pennington

Tired of standing in the yard and holding the garden hose to water the grass? Maybe it's time to buy a rotating lawn sprinkler, invented by Eugene Beggs and A.S. Pennington.

Sand-Grabber—Barbara Derkoski

Barbara Derkoski's first invention, the Sand-Grabber, is a beach umbrella that literally grabs the sand. When you stick the tip into the sand, it acts as an anchor, holding tight even in winds of up to thirty knots. Keep yourself and other sunbathers safe from flying umbrellas with the "Sand-Grabber."

Derkoski, a graduate of Newark's West Side High School, is a former designer, owner, operator and manufacturer of bridal fashions at Lady Barbara Designs in Manahawkin. Today, she is founder and president of BeeTee Enterprises in Tuckerton.

Barbara Derkoski was inducted into the New Jersey Inventors Hall of Fame in 1996.

String Thing—David Brown, 1990s

Seeing that his father was having trouble tying stacks of newspapers, David Brown devised a successful prototype for solving the problem. His String Thing is a small, notched piece of recyclable cardboard with a center hole, slits on the sides and an attached eight-foot-long string. The person tying up newspapers loops the string around the materials, pushes the string through the center hole and ties the bundle tightly. Next, string is wrapped around the materials in the other direction. Finally, the string is pulled into the side slots, completing the process. The entire process should take less than one minute.

After four years of planning, Brown and his aunt, Virginia Brown of South Orange, started Zel Products Inc., based in East Brunswick. As of 2001, the company sold thousands of String Things annually to counties and municipalities. Regular customers include Hudson County, the cities of Newark and Jersey City and the townships of East Brunswick and Egg Harbor. Pittsburgh Steelers football player Josh Miller markets the invention, appearing in uniform on the package with the slogan: "Josh Miller Gets on the Recycling Kick!" The String Thing is Brown's first and only patent. Also named on the String Thing patent are Brown's Aunt Virginia and his father, Douglas.

David Brown was inducted into the New Jersey Inventors Hall of Fame in 2001.

Just for Fun: Toys and Other Stuff

Visible Man and Visible Woman—Marcel Jovine, 1960

Born in Naples, Italy, Marcel Jovine (1921–2003) is known for designing and creating the Visible Man, Visible Woman, Visible Engine and the Blessed Event Doll.

During World War II, Jovine was captured by the Allies. While serving time in Pennsylvania at a POW camp, he enjoyed sketching and creating sculptures. After the war, he returned to Italy but soon came back to the United States to marry Angela D'Oro, whom he had met during the war. The couple settled in New York, where Jovine decided to create toys. His first success, the Blessed Event Doll, was eighteen inches tall, and its face appeared to be crying. The invention was bought by the Ideal Toy Company and became a big seller. The royalties provided Jovine with the money to buy a large Victorian house in Closter, New Jersey.

By the early 1960s, Jovine began to create model kits, one of the most popular being the Visible Man, a clear plastic human figure with its internal organs shown. The cover of the model comes off so that the vital organs can be removed. The Visible Woman is similar but also features an optional womb to simulate pregnancy. Jovine also created the Visible Engine, a working model of a V-8 engine that could be assembled from plastic parts. Visible Man, Visible Woman and other "Visible" models are still on sale today.

Jovine also invented other toys specifically designed for boys, including a pirate ship with a full crew and tanks and weaponry based on declassified army blueprints. He was also known for his lifelike plastic figures illustrating popular fairy tales such as *The Three Little Pigs, Little Red Riding Hood, Jack and the Beanstalk* and *Hansel and Gretel*. Drawing on scientific studies, he also fashioned Cosmorama, a working model of a planetarium. Later in his life, Jovine turned to sculpture, creating figures of racehorses. He also designed medals for the 1987 U.S. Olympic team and a five-dollar gold coin for the U.S. Bicentennial.

Play-Doh—Kay Zufall

What toy can change into any color or shape that you want it to be? It's Play-Doh! And it was the brainchild of a nursery school teacher from Dover, New Jersey. But Play-Doh didn't begin life as a toy, as Joe McVicker and Bill Rhodenbaugh of Cincinnati, Ohio, had originally created the squishy material to clean wallpaper. McVicker, the brother-in-law of Kay Zufall, had a company in Cincinnati that manufactured soap and wallpaper cleaner. In the first half of the twentieth century, many people covered their walls with decorative wallpaper. At that time, most homes were heated by coal. Unfortunately, the soot from the burning coal left a coating on the walls, so the wallpaper had to be cleaned once in a while.

McVicker's company, Kutol Products, made a successful cleaner that removed the soot but didn't harm the paper. It was a non-crumbly mixture of flour, water, salt and borax. The product was very successful until two things happened. After World War II, fewer homes were heated by coal, resulting

in fewer homes having problems with soot. Then, in the 1950s, vinyl wallpaper was invented, and it was easy to clean with just soap and water. People no longer needed McVicker's cleaner, so he and his partner stopped making it.

Play-Doh was first advertised on *Captain Kangaroo, Ding Dong School* and *Romper Room. Courtesy of Janet Crisafi.*

Back in Dover, Mrs. Zufall had seen an article about using the wallpaper cleaner to create art projects. Since it was no longer being manufactured, it was hard to find. But, luckily, the clerk in the hardware store found one old can. The idea worked! The dough was soft and pliable, easy for children's fingers to mold. The children in her nursery school class rolled out the dough, used cookie cutters to make shapes and baked it in the oven. Zufall excitedly called her brother-in-law and suggested that he keep making the wallpaper cleaner but market it as modeling clay instead. She was sure that schools and families would buy it.

McVicker and Rhodenbaugh immediately realized that it was a great idea. They were a bit embarrassed, however, because a few years earlier, a man who worked with wallpaper had shown them animal sculptures that he had modeled from the wallpaper cleaner. The two manufacturers hadn't thought much of the idea at the time. But when Kay showed them, they knew that kids would love to play with this product. The men from Cincinnati dreamed up the name Rainbow Modeling Compound, but the nursery school teacher said that that wasn't a name kids would notice. She and her husband, Bob, brainstormed and thought up a new name: Play-Doh.

But how could they market Play-Doh? How could they let the world know about this great product? Joe McVicker had a brilliant idea. He made a deal with children's television star Bob Keeshan, also known as Captain Kangaroo. If the captain would use Play-Doh on his show once a week, he would receive 2 percent of the money from sales. Two other children's shows, *Ding Dong School* and *Romper Room*, soon began featuring the clay as well, and the word began to spread.

The chemical formula was changed a bit, taking out the kerosene-based oil and substituting mineral oil. Today, Play-Doh is mainly a mixture of water, salt and flour. It does not contain peanuts, peanut oil, milk byproducts or latex, so it is safe for almost everyone. Only children or adults who are allergic to wheat gluten or specific food dyes may have allergic reactions to this product. McVicker and Rhodenbaugh also added an almond scent, reminiscent of cookies. In 1955, Play-Doh was first sold in large cans to school districts. It was later packed in smaller containers, and toy stores and Macy's began to sell it.

Monopoly—Ruth Hoskins, 1929; Charles Darrow, 1933

On January 5, 1904, Lizzie J. Magie of Virginia received a patent for her board game, The Landlord's Game. She had created the game to show that the renting of land profited a few people (the landlords) instead of the majority of people (the tenants). The game demonstrated the unfairness of land monopolies and the use of the land tax. Forming a game company with friends, Lizzie published The Landlord's Game in 1906. In 1910, Parker Brothers published her humorous card game, Mock Trial. Two years later, The Landlord's Game was adapted in Scotland by the Newbie Game Co. as Bre'r Fox and Bre'r Rabbit. Although the instructions claimed it was protected by a British patent, there is no evidence that this was actually done.

In 1924, Lizzie revised The Landlord's Game and received a patent for it. In 1932, she published a revised version of the game. Her company sold her final board-game inventions, Bargain Day and King's Men, in 1937, and a third version of The Landlord's Game in 1939. Parker Brothers agreed to publish her later version of The Landlord's Game, but few copies are known to exist.

In 1929, Ruth Hoskins learned the game of Monopoly in Indianapolis, playing with her brother James and his friend Robert Daggett, who was a friend of Dan Layman. When Ruth began teaching at the Friends School in Atlantic City in October of that year, she taught Monopoly to other teachers, students and Quaker acquaintances. Ruth and her friends made their own Monopoly boards and figured out the rules. Friends of Dan Layman, a college student in Reading, Pennsylvania, introduced him to an early copy of Monopoly during the late 1920s. After leaving college, Layman returned to his home in Indianapolis and decided to market a version of the game, calling it Finance, but it was not yet in stores in the late 1920s and not available on the East Coast until years later.

Since Ruth and her friends all lived in or near Atlantic City, they changed the place names on the board from Indianapolis streets to streets and towns in the Atlantic City area. One housing development, Marven Gardens (note the "en" spelling), was in Ventnor, so they used both of those names in the game. They added four railroads, a "Free Parking" space and a "Go" space to begin the game. Because it was such fun to play, Ruth and her friends introduced the game to many others—among them were the Harveys and the Raifords. The Raifords showed the game

The Monopoly board features streets and towns in the Atlantic City area, four railroads, a "Free Parking" space and a "Go" space. *Courtesy of Robert H. Barth.*

to Charles E. Todd, whose business was restoring old hotels. Todd lived at the Emlen Arms Apartment Hotel and was its managing director. Among the many occasional guests at the hotel were Charles Darrow and his wife. Soon, the Todds, Raifords and Darrows were playing Monopoly together. Darrow, who was unemployed and needed money, asked Todd to write down the rules of the game.

It is thought that Darrow then began making copies of the game and selling them for four dollars each, offering them to department stores in Philadelphia. In time, he had acquired so many orders that he couldn't keep up with production. He then tried to interest the Parker Brothers company in the game. Parker Brothers declined, noting that there were fifty-two errors in the game, among them the fact that it took too long to play and the rules were too complicated. Darrow paid a printer to make five thousand copies and continued to sell the game to stores, including F.A.O. Schwarz in New York City. A friend of the president of Parker Brothers bought the game and suggested that the company reconsider. The president and Darrow came to an agreement: Parker Brothers would buy the game and give Darrow a royalty on each game sold. Darrow permitted the company to make a shorter version of the game. These royalties made Darrow a millionaire, the first game inventor to make that much money. Atlantic City has erected a plaque in his honor on the Boardwalk.

While it appears that Darrow, who claimed the game was his "brainchild," was not the inventor of Monopoly, the game he patented became a bestseller for Parker Brothers. After buying the patent from Darrow, Parker Brothers may have learned that there were other versions of the game. The company then secured patents and copyrights and bought, developed and published The Landlord's Game and Finance. Parker Brothers claims that Charles Darrow was inspired by The Landlord's Game to create a new diversion to entertain himself while he was unemployed.

We will leave it to the reader to decide who was the true inventor of Monopoly.

Coin Tiles—Dianne Leoni, 2012

Hillsborough teacher Dianne Leoni collaborated with ETA hand2mind to produce her invention, Coin Tiles. Noting that, despite technology, children are not getting exposure to using coins, Dianne created a product that helps students identify coins and make change. This game allows students to manipulate parts of a dollar to see a visual relationship between the value of pennies, nickels, dimes and quarters. Through the use of an area model, Coin Tiles give students meaning to the relationship between coins, building an understanding of coin recognition, coin value, coin equivalence and making change.

The Joy Buzzer and Other Practical Jokes—Samuel Sorenson Adams, 1928

Samuel Sorenson Adams, founder of the S.S. Adams Company in Neptune, invented over seven hundred practical jokes and held dozens of patents. The most famous of his patented gags include the Dribble Glass, a glass with hidden slits, causing water to drip on the "victim"; the Joy Buzzer, a ring with a noise-maker that shocks when shaking hands with the prankster; and the Snake in the Jar, a spring wrapped in cloth and squeezed into a can—when the lid comes off, out springs the snake!

As a young man, Adams worked for a dye company. While working there, he discovered that a gray powder, left over from the manufacturing operation, made people sneeze. He began selling it to his friends as a joke. Then, in 1906, Adams began the Cachoo Sneeze Powder Co. Among his earliest inventions were the Rackett Exploding Cigarette Box, Itching Powder and Stink Bombs. He loved squirting tricks, magic tricks, puzzles and party favors.

The Joy Buzzer, created in 1928, was so popular that Adams made enough money to move from his small Asbury Park factory to a much larger building in Neptune. Here he made magic smoke, bugs in plastic ice cubes, exploding cigars, snapping gum, foaming sugar, milk blobs, ink blots and bird poop.

More Toys and Games—Randice-Lisa Altschul, 1985

With little technical training, Randice-Lisa Altschul began creating games and toys for children and adults. By the age of twenty-six, this New Jersey toy inventor was a millionaire, having licensed more than two hundred games and toys. Some of her successful ventures include the Miami Vice board game, based on the popular television show; a Barbie's 30th Birthday game; and board games based on popular cartoons such as *Teenage Mutant Ninja Turtles*, *Power Rangers* and *The Simpsons*. Altschul also created a breakfast cereal that comes in the shape of action figures and dissolves into mush in milk. Altschul was the first inventor represented by the William Morris advertising agency and the first contracted by NBC to develop merchandise based on television programs.

Flexible Flyer Sled—Samuel Leeds Allen, 1889

Born into a famous Philadelphia family, Samuel Leeds Allen (1841–1918) attended Westtown Boarding School, a Quaker academy in Moorestown, New Jersey. Not being a city boy, Allen decided at the age of twenty to learn agriculture at his father's farm near Westfield in Cinnaminson Township. In 1866, Allen married Sarah Roberts, and the couple moved to Ivystone Farm near the village of Westfield, where they would live and work for almost thirty years.

Allen and his father opened a manufacturing company to produce farm equipment, and he was awarded almost three hundred patents for farm machinery, including the fertilizer drill, seed drill, potato digger, cultivator, furrower, pulverizer, grass edger and other implements. Due to his Quaker beliefs, Allen was one of the first company owners to provide modern benefits for his workers. These included a clinic with a nurse, a dining room for hot meals, an auditorium for entertainment programs, death and disability insurance and a retirement plan.

Manufacturing farm equipment was seasonal, however, and Allen needed a new product to keep the workers busy during the summer and fall months. The answer was a sled for coasting down snowy hills. With his children and the students of Westtown School as testers, Allen created many versions of steerable sleds. One, the Fairy Coaster, had double-steel runners and plush seats that could hold three or four adults. Even better, the sled could be folded up and taken on a streetcar or train to the owner's favorite hill. The only problem was the cost: fifty dollars!

Allen designed a smaller version of the Fairy Coaster, but it was too small to steer properly. Not one to be stopped by adversity, Allen finally came up with the perfect steerable sled: the Flexible Flyer. As A.L. Jacoby wrote in an account of the invention:

> *Mr. Allen worked up a sled with only one pair of runners, made of rounded steel, and had these runners weakened at one point about halfway back to form a sort of hinge, so they could be bent sidewise there. This gave the steering effect of double-runner sled, but with continuous runner. This first flexible runner sled was never tried out in snow, but it gave Mr. Allen the right idea, and a sled with flexible T-shaped runners and a slatted seat was soon made. After it was a proven success, it was named by Mr. Allen, the Flexible Flyer.*

On Valentine's Day 1889, Allen applied for a patent for his sled. It was granted six months later. In the beginning, the sled did not sell very well, but after Wanamaker's Department Store in Philadelphia and R.H. Macy's in New York were convinced to sell them, the Flexible Flyers became popular. Before Christmas 1915, Allen noted, "We have been selling sleds at a great pace, averaging 2,000 per day, and the demand is so urgent we are sending whole [rail] car loads of about 1,200 each to New York, New Haven, and Pittsburgh by express; perhaps five cars in all."

Samuel Leeds Allen was inducted into the New Jersey Inventors Hall of Fame in 2004.

Topper Toys—Henry Orenstein, 1951 to the present

Henry Orenstein founded the De Luxe Toy Company in 1951, with its main product being fashion dolls (before the introduction of the Barbie Doll). In the late 1950s, Orenstein sold the company to the Philadelphia & Reading Company, a railroad business, which renamed it De Luxe Reading Toys. Realizing that it knew nothing about the toy business, P&R sold the company back to Orenstein, who renamed it Topper Toys. The company moved from its Newark and Bloomfield locations to the former Singer Sewing Machine plant in Elizabeth, New Jersey. Orenstein is known for inventing dolls whose eyes magnetically opened and closed and dolls whose hair grew with the twist of an arm. For *Sesame Street*, he created "Walking Letters."

At a toy fair in 1983, Orenstein spotted robot figures that changed into vehicles. Working with the Japanese company that owned them, he gave the robots personalities and developed a story line: those that changed into cars were at war with those that transformed into planes. The Hasbro toy company named the toys Transformers and in 1984 sold enough of them to net $115 million.

Among the many toys created by Orenstein are Baby Magic, Battlewagon, Charlie and Me, Ding-A-Ling Robots, Funny Face, Johnny Astro, Lillie Doll, Mr. Pierre, Penny Brite Doll, Dolly Surprise, Silly Safari, Dawn Dolls, the Suzy Homemaker Doll and Johnny Lightning 500 (race cars and track).

In the 1990s, Orenstein learned to play poker, but he did not enjoy watching it on television because viewers could not see the players' cards. His inventive mind went to work, and he developed a table equipped with tiny cameras that would show each player's cards. In 1995, he received a patent for that idea, but the television networks thought that professional poker players would not want their cards shown. Eventually, however, channels like ESPN, Travel, Bravo, the Game Show Network and Fox Sports (on which Orenstein's own poker show airs) began using a variation of his design.

At the age of nineteen, Henry Orenstein was captured by the Nazis during World War II and spent time in five different concentration camps. He came to the United States after the war and became a commercial success, bringing joy to millions of children with his toy creations.

Henry Orenstein was inducted into the New Jersey Inventors Hall of Fame in 2006.

Bottle Pals—Bonnie L. James, 1989

While feeding two-month-old Jessica Lee in 1989, Bonnie L. James of Tuckerton conceived the idea for Bottle Pals, baby-bottle holders that are amusing, educational and appealing to both mothers and children and, more importantly, that make feeding babies easy and fun. Bottle Pals, brightly colored stuffed animals with holes in their middles to hold a baby bottle, support the bottle while simultaneously amusing the infant. In addition, they help promote hand-eye coordination, relaxation and better digestion for the baby. Currently, Bottle Pals are sold at J.C. Penney, Toys "R" Us and Child World, among other retailers. Bottle Pals are available in four designs: bear, duck, dog and bunny.

Bonnie L. James was inducted into the New Jersey Inventors Hall of Fame in 1993.

Dolls and Toys—Barbara Yaney

Barbara Yaney of Somerset has spent decades creating dolls and toys for clients such as Harley-Davidson, Disney and McDonald's. Among her creations are the Celine Dion collection, Curious George for Russ Berrie, Rugrats for MTV/Nickelodeon, Sam the Eagle for the U.S. Olympics, Avon dolls and the Yellow Box Collection for Hallmark.

Toy Locomotive—Eugene Beggs

Born in 1836, Eugene Beggs was a pioneer in creating miniature railroads for children. He built the first alcohol-fueled, steam-powered miniature toy locomotive. Beggs died in 1924 and is buried in Cedar Lawn Cemetery in Paterson, New Jersey.

Model Railroad Rail Joiner—Stephan Schaffan Jr., 1949

As a boy, Stephan Schaffan built model airplanes and often visited the local hobby shop. Looking for a job, he asked the owner what he could do. "Here," said the man, giving him some train track, "see if you can improve this." Schaffan soon created the first "switch kit." When they became bestsellers, he and his family built more in the basement at night. Soon, Schaffan was improving the connections that held two pieces of track together. Instead of soldering the pieces (a time-consuming operation), hobbyists could now simply stick the pieces of track together. Schaffan's invention of the first practical rail joiner, along with his switch kits, led him and his father to open their first factory at 413 Florence Avenue in Hillside, New Jersey.

Today, the Atlas Model Railroad Co. Inc. sells about 5 million pieces of HO and N-scale model track each year. The company also sells model trains, buildings, electrical components and miniature people and animals for railroad layouts. Atlas's sister company, launched in 1998, manufactures and sells O-gauge track, freight cars and building kits. Schaffan also patented other inventions, including the Super-Flex Track, Snap-Switches and Custom-Line turnouts.

Stephan Schaffan was inducted into the New Jersey Inventors Hall of Fame in 2000.

Vibrating Bed—John Houghtaling, 1958

Tinkering in the basement of his Glen Rock, New Jersey home, John Joseph Houghtaling (pronounced HUFF-tay-ling) invented the Magic Fingers vibrating bed, found in many hotels in the 1960s and 1970s. Attached to a bed, the device brought fifteen minutes of "tingling relaxation and ease" for only twenty-five cents, according to the advertisements.

Houghtaling (1916–2009) had been selling beds with a built-in vibrating mechanism when he realized that it would be cheaper to create something that would attach to the outside of an existing bed. In its most popular years, the gadgets collected about $6,000–$7,000 a month in quarters. After his retirement, Houghtaling continued to invent and sell coin-operated machines such as scales and pulse-checking devices.

PERSONAL HEALTHCARE

Dental and Cosmetic Products—Anthony E. Winston

Need to whiten your teeth, get rid of plaque or prevent cavities? Some of Anthony Winston's inventions may help you. While working for Church & Dwight, the manufacturers of Arm & Hammer products, Winston received thirteen patents for using baking soda to create dental items such as pastes, gels and tartar-control agents. These inventions are effective in removing plaque, whitening teeth and controlling gingivitis. Other creations also strengthen tooth enamel, prevent cavities and repair teeth damaged by acid. The latter, earning fifteen patents, combines calcium and ions of phosphate and fluoride that are applied to the teeth by means of chewing gums, pastes and mouth rinses, replenishing needed minerals.

In addition to dental items, Winston has also made products that help farmers kill harmful fungi on plants. These products are environmentally safe, and the fungi have not been able to develop a resistance to them. Winston also created a deodorizer, an earwax removal aid, metal cleaners and laundry detergents, earning ninety-five patents in all. He co-authored the 1996 book *Handbook of Aqueous Cleaning Technology for Electronic Assemblies*.

Anthony Winston was inducted into the New Jersey Inventors Hall of Fame in 2002.

Sunscreen—Victor Palinczar, 1986

Beginning in 1983, Victor Palinczar worked in his Trenton laboratory to develop a waterproof sunscreen that would protect people from the harmful rays of the sun and not be washed off by water or perspiration. In that same year, he founded Princeton Products Research, a company that develops and licenses new products to the health and beauty industry. This invention was the forerunner to the high-SPF waterproof sunscreen products on the market today. It helps prevent aging of the skin, wrinkling and skin cancer. Palinczar has been awarded twenty-two patents worldwide and has launched fifty-two products domestically.

Victor Palinczar was inducted into the New Jersey Inventors Hall of Fame in 1992.

Antibacterial Toothpaste—Abdul Gaffar, 1988

Working at the Colgate-Palmolive Company in Piscataway in 1988, Princeton resident Abdul Gaffar developed Colgate Total, the nation's first antibacterial toothpaste. Some say that it is the most important breakthrough in dental health since fluoride was introduced in the 1950s. Gaffar received four patents that contributed to the toothpaste's discovery and success.

Abdul Gaffar was inducted into the New Jersey Inventors Hall of Fame in 2001.

Solar-Powered Water Pumps—Quentin T. Kelly, 1992

Quentin T. Kelly founded his WorldWater Corporation in 1984 to develop solar-powered pumps and power, mainly for use in poor countries. Much of the early work was done by Princeton University engineers, retired managers from the United Nations Children's Fund (UNICEF) and volunteers.

In developing nations, people spend hours carrying water from sources that are often unsafe. Kelly's solar pump enables these people to obtain clean water from their own deep wells, removing the need for the women and girls to walk five to eight kilometers every day to fetch water from a surface pond or lake, which is generally contaminated (animals and people share such watering holes for bathing as well as drinking). When they are given solar-powered pumps, the people can get safe drinking water from distant places without the expense of gasoline or diesel generators.

WorldWater pumps are distributed to villages in Malawi, Mozambique, Tanzania, Uganda, South Africa and the Philippines. WorldWater has improved on its earlier systems and can now pump from contaminated surface sources, rivers, lakes and shallow wells, providing up to thirty thousand gallons of purified drinking water per day.

Quentin T. Kelly was inducted into the New Jersey Inventors Hall of Fame in 1998.

CHAPTER 5

BRIGHT IDEAS:
SCIENTIFIC AND TECHNOLOGICAL
BREAKTHROUGHS

Theory of Relativity—Albert Einstein, 1915

Albert Einstein (1879–1955), the famous scientist who spent the last twenty-two years of his life in Princeton, New Jersey, proposed the theory of relativity in 1915. Although he did not introduce the concept of relativity, he recognized that the speed of light in a vacuum is constant and is an absolute physical boundary for motion. This theory does not impact our lives, since we travel at speeds much slower than light. It does, however, affect objects traveling at or near light speed, as it states that objects will move slower and shorten in length from the point of view of someone on Earth. Einstein also produced the $E=mc^2$ equation, which shows the equivalence of mass and energy.

The first part of Einstein's theory, the special theory of relativity, explains that rest and motion are relative. Part two, the general theory of relativity, shows that objects continue to move in a straight line in space-time but that we observe the motion as acceleration due to the curved nature of space-time. This explained the phenomena of light bending around the sun and predicted black holes. This general theory demonstrates that time is linked to matter and space; thus, time, space and matter constitute what we call a continuum—the three must come into being at the same instant. Time cannot exist without matter and space.

For his work on relativity, the photoelectric effect and blackbody radiation, Einstein was awarded the Nobel Prize in 1921. Among his other theories

Albert Einstein. *Library of Congress.*

and discoveries are the law of photoelectric effect, the gas theory, the derivation of Planck's radiation law, the law of photochemical reactions and the theory of related heats. His exploration of the principles of the process of stimulated emission of radiation led to the development of the first laser. Einstein also held patents for a noiseless household refrigerator, a hearing aid and a camera.

In 1933, Einstein denounced his German citizenship for political reasons and immigrated to the United States to become a professor of theoretical physics at Princeton's Institute for Advanced Study. He became an American citizen in 1940 and retired from his position at the university in 1945. After he retired, Einstein lobbied for nuclear disarmament and a world government. Along with Albert Schweitzer and Bertrand Russell, he fought against nuclear tests and bombs. As his last public act, and just days before his death, he signed the Russell-Einstein Manifesto, which led to the Pugwash Conferences on Science and World Affairs. Einstein's latter years were also spent searching for a unified field theory, a universal force that would link gravitation with electromagnetic and subatomic forces. He died at Princeton in 1955.

Albert Einstein was inducted into the New Jersey Inventors Hall of Fame in 1989.

VOLTMETER, AMMETER AND WATTMETER— EDWARD WESTON, 1886

A prolific inventor who held 334 patents, Edward Weston helped revolutionize the measurement of electricity. In 1886, he developed a practical, portable instrument to accurately measure electrical current. This device became the basis for the voltmeter, ammeter and wattmeter. His company, Weston Instruments, also manufactured ohmmeters, transducers and transformers, as well as other precision instruments, arc-lighting systems, the magnetic speedometer and the dashboard ammeter for Harley-Davidson motorcycles.

Edward Weston was inducted into the New Jersey Inventors Hall of Fame in 1990.

Sensors for Heating and Cooling Systems and the Gas Density Switch—Beatrice Alice Hicks, 1946–'50s

Born in Orange, New Jersey, Beatrice Alice Hicks (1919–79) became the first female engineer employed by Western Electric, a subsidiary of Bell Telephone, in 1942. (She and many other women had this opportunity in a male-dominated industry because so many men had left to fight in World War II.) When her father died in 1946, Hicks took over his business, the Newark Controls Company. Over the years, she developed environmental sensors for heating and cooling systems. In addition, she invented a gas density switch, an engineering breakthrough that was necessary for space travel. The switch signaled when the atmosphere changed inside a closed environment (for example, when there was a leak). The switch could also open a valve or start a pump. It was used in the ignition system on the Saturn V rockets.

Transistor—John Bardeen and Walter Brattain, Bell Labs, 1947; William Shockley, 1947

On December 16, 1947, while working at Bell Labs, John Bardeen and Walter Brattain created the transistor, a small electronic device that amplifies a weak input signal. The team had rigged up thin slices of silicon and germanium with a wire tipped by a gold-foil point and was able to show that the contraption could act as an amplifier. Refinements of the transistor have gradually replaced virtually all uses of vacuum tubes. Immediately, the two men submitted a patent application for a working point-contact transistor. Soon, their boss, William Shockley, went to work trying to improve this first transistor. He made changes that made the device easier to use and manufacture, and he applied for another patent nine days later.

Today's transistors are tiny—up to a billion can be used in a single microchip. Transistors power satellites, fiberoptic networks, radios, computers, cell phones and fax machines. Using transistors, these devices can send words, numbers and images around the globe. In his

While working at Bell Labs in Murray Hill, New Jersey, John Bardeen and Walter Brattain created the transistor. *Reprinted with permission of Alcatel-Lucent USA Inc.*

2012 book *The Idea Factory*, Jon Gertner notes, "A recent Intel computer processor chip not much larger than a postage stamp contains two billion transistors. (Intel, moreover, manufactures about 10 billion transistors every second.)"

John Bardeen, Walter Brattain and William Shockley received the Nobel Prize in 1956. All three men were inducted into the New Jersey Inventors Hall of Fame in 1994.

SILICON-GERMANIUM THERMOELECTRIC POWER GENERATOR—BENJAMIN ABELES AND GEORGE CODY, 1960S

In 1977 and 1979, the National Aeronautics and Space Administration (NASA) launched two space probes to explore the outer planets. This exploration has been described as "an epic journey to the four giant gas planets of our solar system." The invention that made the Voyager missions possible, the silicon-germanium thermoelectric power generator, was developed in the 1960s by Benjamin Abeles and George Cody at the David Sarnoff Research Center in Princeton when the two scientists discovered low thermal conductivity in the silicon-germanium alloys they were using. They needed low temperatures in order to optimize the electrical properties required to efficiently convert heat to electricity. This discovery led to the development of a reliable and long-lasting source of energy, perfect for deep space probes that cannot use solar panels to collect energy. In 1980, the Franklin Institute presented the researchers with the prestigious Stuart Ballantine Medal for advancing communications using thermomagnetic radiation.

Benjamin Abeles and George Cody were inducted into the New Jersey Inventors Hall of Fame in 1991.

Solar Panels—Gerald Pearson, Calvin Fuller and Daryl Chapin, 1955

It is thought that New Jersey homes and businesses use more solar panels than any state other than California. But did you know that the panels were invented in the Garden State in 1955? In that year, Gerald Pearson, Calvin Fuller and Daryl Chapin designed the first photovoltaic cell at Bell Labs. The first trial of their panel, made of razor blade–sized strips of silicon, occurred at a telephone company in Georgia. In 1958, Bell Labs supplied solar electric power for NASA's first permanent satellite. Following this initiative, New Jersey became the site of the largest solar research and development effort in the country when Exxon made a large-scale investment at its Linden and Florham Park facilities.

Gerald Pearson, Calvin Fuller, and Daryl Chapin were inducted into the New Jersey Inventors Hall of Fame in 2006.

In 1955, Gerald Pearson, Calvin Fuller and Daryl Chapin designed the first photovoltaic cell (solar panel) at Bell Labs in Murray Hill, New Jersey. *Reprinted with permission of Alcatel-Lucent USA Inc.*

MASER—BELL LABS, 1957

The maser, an acronym for **M**icrowave (or Molecular) **A**mplification by **S**timulated **E**mission of **R**adiation, is a device that produces coherent electromagnetic waves through amplification by stimulated emission. In 1957, when the optical coherent oscillator was first developed, it was referred to as an optical maser, or laser (**L**ight **A**mplification by **S**timulated **E**mission of **R**adiation). This acronym was created by Gordon Gould in 1957.

Laser—Arthur Schawlow, 1958

The invention of the maser led Arthur Schawlow to apply the same principles to light amplification. The helium-neon laser, the first laser that could emit a continuous beam of light, revolutionized the communications industry and the processing of materials, optical scanning, medicine, energy research and surveying. Bell Labs has since developed hundreds of types of lasers for use in science, industry and telecommunications. In about ten years, Bell Labs was able to shrink the device to the size of a grain of salt.

UNIX Operating System—Ken Thompson and Dennis Ritchie, 1969

In the 1960s, Ken Thompson and Dennis Ritchie began working on the framework of a new computer operating system at Bell Labs. They called the system Unix (officially trademarked as UNIX), and it became the language of the personal computer and the basis of everything from supercomputers to the Internet. In 2011, Thompson and Ritchie were awarded the Japan Prize, which honors outstanding achievements in science and technology.

Ken Thompson and Dennis Ritchie were inducted into the New Jersey Inventors Hall of Fame in 2006.

Charge-Coupled Device—George E. Smith and William S. Boyle, 1969

The charge-coupled device (CCD) is a sensor used in many of today's technologies, including bar-code scanners, digital cameras and video cameras. The CCD captures light, converts it to an electrical charge and then moves that charge to the edge of the device, where it can be processed electronically. It has been called the electronic equivalent of a bucket brigade. For this reason, a CCD is often considered the digital version of film.

The CCD is a light-sensitive electronic sensor that uses the responses of electrons to different amounts of light to create photographs and images of extraordinary detail. It moves an electrical charge, usually from within the device, to an area where the charge can be manipulated—for example, conversion into a digital value. Although CCDs are not the only technology to allow for light detection, CCD image sensors are widely used in professional, medical and scientific applications in which high-quality image data is required. In applications where a somewhat lower quality can be tolerated, such as webcams, cheaper active-pixel sensors are generally used.

According to George E. Smith of Waretown, the CCD was the foundation for advances in many electronic fields. Described by Smith as "very simple," the device was instrumental in the creation of digital cameras, semiconductors, capacitors and electron transfers. At Bell Labs, the CCD was initially contained in a solid-state picture-phone camera and a commercial television chip. It has also been useful in scanners, fax machines, medical imaging, bar-code readers, satellite surveillance and astronomy. In 2009, Willard S. Boyle and George E. Smith received the Nobel Prize for Physics for the invention of the CCD sensor.

Willard S. Boyle and George E. Smith were inducted into the New Jersey Inventors Hall of Fame in 2008.

SPECTRA—Sheldon Karesh and Dusan Prevorsek

Sheldon Karesh and Dusan Prevorsek developed the revolutionary, high-strength SPECTRA, a polyethylene fiber that has many applications. SPECTRA is used to make cut-resistant gloves for surgeons; bullet-proof vests for police and military; helmets, armor and armored vehicles for law enforcement agencies; sports equipment such as kayaks, canoes, bicycles, boats, sails, fishing lines; ski ropes; and artificial tendons, ligaments and joint prostheses. SPECTRA is one of the world's strongest and lightest fibers. A bright white polyethylene, it is fifteen times stronger than steel and more durable than polyester. Light enough to float, it also exhibits high resistance to chemicals, water and ultraviolet light.

Sheldon Karesh and Dusan Prevorsek were inducted into the New Jersey Inventors Hall of Fame in 1989.

VIDEO SIGNAL INTERPOLATION USING MOTION ESTIMATION—ARUN NETRAVALI, 1983

A resident of Westfield, New Jersey, Dr. Arun Netravali was the president of Lucent Technologies Bell Laboratories from 1999 to 2001. His patented "Video Signal Interpolation using Motion Estimation" improved high-definition television (HDTV) and benefitted the delivery of broadcast television, compact discs, digital video displays and the Internet. The principles he used are the basis for coding and decoding digital video signals. He holds more than seventy patents in computers as well as digital video technology and received a 1994 Emmy award for his work with the HDTV Grand Alliance.

Arun Netravali was inducted into the New Jersey Inventors Hall of Fame in 2001.

BIRD-SAFE WIND TURBINES—RAYMOND GREEN, 2012

The blades of modern wind turbines can kill birds and bats, which often fly into them. Raymond Green's Catching Wind Compressed Air Enclosed Wind Turbine has no external blades and is safe for flying creatures. Its patented inner-compression cone technology squeezes incoming air as it is drawn through the turbine. This creates increased power output with virtually no noise. The smaller blades are located inside the device, eliminating the sound that traditional blades make as they spin. Lynch's company, Sigma Design of Middlesex, New Jersey, is testing, improving and manufacturing the invention.

RADAR—COLONEL WILLIAM R. BLAIR, 1941–45

Colonel William R. Blair is known as the father of radar. During World War I, he worked with the U.S. Army's Signal Corps and developed his radar theory. Radar is the method by which radio waves can accurately and quickly detect the position of distant "invisible" objects (such as planes). During World War II, Blair's theory became reality when he invented radar. During wartime, however, the invention had to be kept secret, and Blair was not allowed to apply for a patent.

When the war ended in 1945, Blair applied for a patent, but it was challenged by others who claimed to have been the inventors. Finally, after a special act of Congress and extensive research, he was awarded the patent in 1957. The U.S. Army has said that the invention of radar was "as important and far-reaching in its military application as the first U.S. patent issued on the telephone was to commercial communications."

William R. Blair was inducted into the New Jersey Inventors Hall of Fame in 2004.

HEDGEROW CUTTER—CURTIS CULIN, 1944

During World War II, the Allies (United States, Great Britain, Canada, Australia, New Zealand and other countries) invaded the beaches of Normandy, France, in an operation known as D-Day beginning on June 6, 1944. After securing the beaches, the Allies tried to move inland to defeat the German army, which was occupying France. One problem that they encountered was the rows of hedges that grew along all of the roads. Fifteen feet tall, these hedgerows gave protection to the enemy and even stopped the powerful Sherman tanks. But Sergeant Curtis Culin from Cranford, New Jersey, had an idea. He gathered up the iron obstacles that the Germans had placed on the beaches and welded them to the front of the tanks. Like sharpened tusks, these hedgerow cutters tore through the countryside. General Omar Bradley was so impressed that he ordered the invention attached to two hundred tanks, speeding the advance of the Allies and saving thousands of lives.

The Hedgerow Cutter, also known as the Rhino, continued to be used in the Korean and Vietnam Wars.

ROBOTIC TANK CLEANER—HARRY ROMAN, 1988–93

Along the highway, you may have seen large tanks that hold chemicals, petroleum, natural gas and other materials. Years ago, when it came time to clean these tanks, highly trained technicians wearing bulky, non-permeable clothing and breathing apparatus had to be lowered into the structure after it had been drained. This was a time-consuming and dangerous job.

Working for Public Service Electric & Gas (PSE&G), Harry Roman led the Applied Robotic Testing unit. He and his team developed twenty-three robots to inspect tanks, ensuring a safer, more thorough inspection. The tank does not have to be drained, as the robot is lowered through an opening in the roof.

Harry Roman is a strong supporter of invention in New Jersey. He gives lectures about inventors and innovations to students and community groups. He created the Student Project Team program for New Jersey engineering students and has published and/or presented over four hundred papers, books and articles and over six hundred poems, short stories and other works.

Harry Roman was inducted into the New Jersey Inventors Hall of Fame in 2005.

FIRST COMMERCIAL ELECTRON MICROSCOPE— JAMES HILLIER, 1941–53

James Hillier (1915–2007), working with fellow student Albert Prebus at the University of Toronto (Ontario, Canada), built the world's first practical electron microscope. This new creation could magnify an object up to 7,000 times, much greater than the optical microscopes of the time. Instead of using a beam of light, as in traditional microscopes, this new machine passed a beam of electrons through a specimen. The beam was then focused on a photographic plate. Since the wavelength of electrons is smaller than that of light, the microscope had greater magnification and depth of focus, producing very clear images. In 1941, Hillier began working for RCA in Camden, New Jersey. His mission at RCA was to work with a team to build the first commercial electron microscope. He quickly noted that "the real name of the game was to find out how to put significant things" like blood cells and bacteria into the microscope without "burning them to a crisp" in the potent electron beam. RCA's microscopes were used in 1949 at the Sloan-Kettering Institute to view cancer cells taken from animal tumors. By that time, Hillier and his team had increased the magnification to 200,000 times. About two thousand electron microscopes were sold between 1940 and the 1960s, when RCA ended production.

James Hillier was inducted into the New Jersey Inventors Hall of Fame in 1992.

Liquid Crystal Displays (LCDs)—Richard Williams, 1963

Today, liquid crystal displays (LCDs) are used in hundreds of products, including watches, calculators, computer terminals and other electronic devices. Digital readouts are so common that we often don't even think of them as an invention. But LCDs were developed and patented in the 1960s at the David Sarnoff Research Center in Princeton, New Jersey.

Liquid crystals are organic compounds with the appearance and properties of a liquid, but their molecules tend to form orderly arrays like those found in solid crystals. They can be made opalescent and hence reflective by applying an electric charge. In the early 1960s, only laboratory researchers were aware of liquid crystals.

In 1962, Richard Williams filed a patent disclosure on "electro-optic elements utilizing an organic nematic compound," more commonly known as liquid crystals. The patent was granted in 1967. His landmark paper, "Domains in Liquid Crystals," was published in 1963.

Richard Williams was inducted in the New Jersey Inventors Hall of Fame in 1994.

Wattvision—Savraj Singh, 2011

Wattvision makes sensors and software to help consumers and businesses save money on energy. The user can view the whole-house live energy use data, including the hourly cost, on the web or on a mobile device, updated every ten seconds. The homeowner can track usage on a graph and get inside information and never be surprised by an energy bill again! The consumer purchases the equipment, which includes a sensor and a gateway, a small box that is keyed to the wireless network in the home. The sensor is attached to the electric meter and connected by wire to the gateway. The user can view online the real-time energy usage. When an energy spike is detected, the consumer is alerted.

CHAPTER 6
TAKE ME OUT TO THE BALL GAME AND MORE: SPORTS

You probably like sports or know people who do. Maybe you play or watch baseball. Well, baseball wouldn't be the game we know today if it weren't for the Garden State. You may have heard that baseball was invented by Abner Doubleday in Cooperstown, New York. It's time to put that story to rest.

First Baseball Game with Modern Rules— Alexander Cartwright, 1846

Ask people where the first baseball game was played, and they will most likely say, "Cooperstown." They may even tell you that the game was invented by Abner Doubleday. It's a good story, but it's not the truth. In fact, the rules of the game and the modern baseball field were created in 1845 by New Yorker Alexander Cartwright (1820–92). While working at the Knickerbocker Fire Station, Cartwright played "town ball" (an early version of baseball) on a vacant lot in Manhattan. When the team could no longer use the lot, Cartwright moved the players across the Hudson River to Elysian Field in Hoboken, New Jersey. Unfortunately, it cost seventy-five dollars to rent the field, so Alexander had to find a way to raise the money. Then he had an idea! He organized a ball club in September 1845 so that

The first game of baseball with modern rules was played in Hoboken in 1846. *Library of Congress.*

he could charge admission and pay for the rental of the field. The club was called the New York Knickerbockers, in honor of the fire station where Cartwright worked. When he formed the Knickerbockers, Cartwright also created a formal set of rules that each player had to agree to. These rules may have been similar to those in earlier versions of baseball, but they are the basis of modern baseball. Cartwright proposed that the distance between all bases should be equal, that a batter should get three strikes before he can be called out and that there should be three outs in an inning. He also proposed the addition of an umpire and the creation of fair and foul territories and eliminated the practice of throwing the ball at the baserunner.

Some of the players did not want to leave their homes in New York to practice in New Jersey, so they stayed in Manhattan and formed their own club, the New York Nine. On June 19, 1846, the New York Nine played the New York Knickerbockers at Elysian Field in Hoboken. It was the first baseball game between two different teams. The Knickerbockers lost 23–1 in four innings. Cartwright was the umpire and fined one player six cents for cursing.

And speaking of Abner Doubleday, he did live in New Jersey (in Mendham) after the Civil War. In 1839, when he is supposed to have been in Cooperstown, he was attending college at the United States Military Academy at West Point. He graduated in 1842 with a commission in the artillery and served in the Mexican War and the Civil War. He was also stationed in Charleston Harbor when the Confederacy attacked Fort Sumter, South Carolina, in 1861. The story that Doubleday invented baseball is almost certainly untrue. He never referred to the game, nor did he ever claim to have invented it. His obituary in the *New York Times* did not mention baseball, either.

BASEBALL RUBBING MUD— RUSSELL "LENA" BLACKBURNE, 1938

Lena Blackburne (1886–1968) began his Major League baseball career as an outfielder with the Chicago White Sox and later became the team's manager. While working as a third-base coach for the old Philadelphia Athletics later in his career, Blackburne heard lots of complaints from umpires. But one in particular would change his life.

During one game, an umpire was unhappy with the condition of the baseballs used by the American League. Since new balls were too shiny, they were rubbed with mud—regular, old-fashioned mud made with water and dirt straight from the field. The trouble was that the mud made the cover too soft, allowing for possible tampering. Something was needed to remove the shine without softening the cover. After the game, Blackburne returned home to Palmyra, New Jersey, and checked out the streams along the Delaware River. He found some muck with a texture that he thought would do the job. In the Athletics' field house, he rubbed some balls with the stuff, and voilà, it worked! Plus, it had no odor and didn't turn the covers black. The umpires were happy, and Lena Blackburne was in the mud-supply business.

The news of the baseball rubbing mud spread, and by the 1950s, every Major League team and some in the minors were using the mud. On his death, the secret location and the mud business were willed to his close friend, John Haas. Haas later turned the business over to his son-in-law, Burns Bintliff, who later passed it on to his son Jim.

Every July, the Lena Blackburne crew harvests hundreds of pounds of "Magic Mud," enough for one season. The mud cures in barrels until the following spring, when it's shipped to teams all over the country. Other kinds of mud and even mechanical methods have been tried to de-slick baseballs, but they couldn't make the grade.

As they say at Lena Blackburne, when the umpire yells, "Play ball," you know that good New Jersey mud will be part of the game.

CATCHER'S MASK—JAMES EDWARD JOHNSTONE, 1922

Baseball's first catcher's mask is said to have been invented in New Jersey by James Edward Johnstone of Newark. Born in Ireland, Johnstone was a minor league pitcher from 1894 to 1897 and a professional umpire from 1902 to 1915. During that time, catchers used a wire-frame mask that wasn't very protective and did not prevent a player from getting a broken nose or losing a tooth.

In 1922, James Johnstone improved on the catcher's mask that had been invented at Harvard.* He developed a new full-vision mask. The frame was a one-piece, solid aluminum casting with horizontal crossbars instead of soldered mesh. Called the Original Full-Vision Mask, it was distributed by the Johnstone Baseball Mask Company, located on Central Avenue in Newark. Johnstone's mask was adopted immediately by Major League catchers, and the original design is still in use today.

James Johnstone was inducted into the New Jersey Inventors Hall of Fame in 1994.

*Harvard baseball player James Tyng became the first catcher to use a mask in regular-season play. To protect Tyng from the newly developed curve ball, team captain Fred Thayer designed a mask similar to those used in fencing. It had head and chin padding and used the mesh of the fencing mask. Tyng made fewer errors, and Thayer patented the mask in 1878. That year, it went into the A.G. Spalding Company catalog. This adapted fencing mask was widely used in the 1880s. According to the August 2004 issue of *Harvard Magazine*, Thayer's mask was made of wire mesh with large eyeholes and cushioning for the chin and forehead.

BATTING CAGE—WELLINGTON TITUS, 1907

Yet another kind of baseball protection was created by Wellington Titus of Hopewell, New Jersey—the batting cage. In 1907, in a section of Hopewell once called Marshall's Corner, Wellington Stockton Titus (1872–1941), known to friends and family as "Welling," made baseball history. As a catcher for the local amateur baseball team, the Hopewell Athletic Club, Titus invented and patented what he called a "base ball back stop." Today, baseball fans know it as the batting cage.

As the story goes, Titus disliked chasing wild pitches and foul tips. To save time, he created a portable batting cage. His cage served as the prototype from which the current batting cages have evolved. The device was an immediate hit, and before the patent was approved, Titus had signed an agreement with A.G. Spalding and Brothers Company to manufacture his portable batting cage. Spalding paid Titus five dollars for each cage sold. The cage was a hit because it was portable and adaptable to use both in and out of doors—and it prevented lost or stolen balls. Prior to Titus's invention, baseball teams hired young neighborhood boys as ball chasers.

When Titus wasn't inventing, he made his living moving houses. His unconventional house-moving methods were said to amaze experts, and it was not unusual for engineering students at nearby Princeton University to watch his productions. To move a house, Titus would often hitch a horse to a beam, which, in turn, was connected to a windlass, a device used for hoisting or hauling. Six to eight men would then place heavy wooden runners under the raised house while six other men soaped the runners to make the building slide. In later years, crankcase drainings were added to the soap to make the house slide even more easily.

Although Titus had never received a formal engineering education, Hopewell residents considered him a natural-born civil engineer. Titus also designed a baseball bat called the "Black Diamond," knitting needles and bootjacks, each one of which featured the head of a different creature of nature. A local foundry molded these unique products.

First World Series Radio Broadcast, 1921

Yet another baseball first for the Garden State! Although the radio had been invented in 1895, it took twenty-six years before the first baseball game was broadcast. On August 5, 1921, on Pittsburgh's KDKA, Harold Arlin called the world's first baseball game broadcast. If this was in Pennsylvania, you may ask, what is the New Jersey connection? In the fall of that year, KDKA installed a connection between Pittsburgh and New York. Famed sportswriter Grantland Rice used this wire to report from the World Series game between the Giants and Yankees. Station WJZ of Newark broadcast the World Series by relay. Sandy Hunt of the *Newark Sunday Call* reported the play-by-play from the Polo Grounds, and Tommy Cowen repeated Hunt's words for listeners.

Golf Tee—William Lowell, 1924

Dentist William Lowell (1862–1954) began playing golf in 1921 at the Maplewood Country Club. He noticed that before driving their first shot, players would often wet their hands and make a small mound out of sand on which to place the ball. Not caring for this method, Dr. Lowell used his dental tools to whittle a golf tee from the wood of a flagpole. On the top of the tee he affixed a cup made of gutta-percha, a tough, rubber-like substance. Lowell made the top of the tee concave so that the ball would rest easily. The dentist asked a woodturner to copy the design.

Although his companions ridiculed the invention, his sons realized its marketing potential. Even professional golfers thought it a silly idea and wouldn't even accept the tees as gifts. Finally, Lowell paid $1,500 to Walter Hagen, the best-known professional golfer of the era, and Joe Kirkwood, a trick-shot artist, to use the tees on their 1922 exhibition tour. Wherever the two went, they left behind tees. Other golfers would then visit their pro shops, demanding the new tees.

Never thinking that golfers would save the tees, Lowell made only five thousand at first, all of them green. Later, he changed the color to red and called them "Reddy Tees." Marketed through the doctor's Nieblo Manufacturing Company, the tees were made of white birch by a woodturner in Norway, Maine.

The golf tee invented by Dr. William Lowell has a concave top that allows the ball to rest easily. *Courtesy of Robert H. Barth.*

Finally, after making numerous improvements, Lowell patented the tee in 1924. He made a small fortune but found that his patent was not specific enough to protect the invention. He spent $140,000 on lawsuits to fight the patent infringements of other inventors.

Ask to see the tees when you visit the U.S. Golf Association Museum (www.usgamuseum.com) in Far Hills, New Jersey.

William Lowell was inducted into the New Jersey Inventors Hall of Fame in 1998.

Note: In 1899, George Franklin Grant, an African American dentist from Boston, patented a wooden tee but did not market it. It was available only to his friends and golfing buddies. Additionally, the top of Grant's tee was not concave; the ball had to be balanced carefully on top.

National Marbles Tournament, Wildwood, 1923

What sport uses even smaller balls than baseball? Marbles! Since 1923, New Jersey has been home to the National Marbles Tournament. In fact, you can visit the National Marbles Hall of Fame in Wildwood. Each June, the annual National Marbles Tournament is held at Wildwood Avenue and the beach. Come and cheer on the seven- to fourteen-year-old "mibsters" (marble-shooters) in this four-day event. You can check out www.nationalmarblestournament.org or contact the Wildwood Historical Society at 609-523-0277.

Mibsters, or marble players, prepare for the annual tournament on the beach in Wildwood. *Courtesy of the Greater Wildwoods Tourism Authority, Wildwood, New Jersey.*

FIRST INTERCOLLEGIATE FOOTBALL GAME, NEW BRUNSWICK, 1869

New Jersey has a famous football connection, too. The first intercollegiate football game was played at College Field in New Brunswick on November 6, 1869. Rutgers beat visiting Princeton University 6–4.

FIRST INDOOR COLLEGE FOOTBALL BOWL GAME, ATLANTIC CITY, 1964

The first indoor college football bowl game was played in Atlantic City in 1964. This game, played on December 19, featured the University of West Virginia and the University of Utah.

FIRST PROFESSIONAL BASKETBALL GAME— TRENTON, NOVEMBER 7, 1896

Played at the Trenton Masonic temple, the first professional basketball game featured the Trenton YMCA and the Brooklyn YMCA. An admission fee was charged, and each player received fifteen dollars (except for Fred Cooper, who got sixteen dollars because he was also the coach). Trenton defeated Brooklyn 15–1.

ICEMAT ICE-RINK FLOORS—CALVIN MACCRACKEN

In 2010, Calvin MacCracken (1920–99) was posthumously awarded the Frank J. Zamboni Award for his development of many products for ice arenas around the world, including his IceMat ice-rink floors. The IceMat creates perfectly uniform ice with much less need for pumping power. The

IceMat systems are prefabricated, portable and permanent and can be used indoors or outdoors. They cost less than ordinary ice rinks and are easier to construct. The IceMat can be installed in a matter of days and disassembled for storage just as quickly. Additionally, the *Hockey News* reported that the IceMat provides the best ice surface in the National Hockey League. MacCracken's invention can be used for temporary or permanent ice skating rinks, skating and curling rinks, luge tracks, toboggan and ski runs and speed-skating ovals.

Calvin D. MacCracken was inducted into the New Jersey Inventors Hall of Fame in 1989.

CHAPTER 7

ON THE MOVE: TRANSPORTATION

When you and your family travel around the neighborhood, the town or the state, you probably use the family car. But years ago, before cars were invented, folks had to ride a horse, sit in a wagon or walk. Since people have always wanted to get places fast, clever New Jerseyans have created quite a few innovations in transportation.

AVIATION

First U.S. Balloon Flight—Jean-Pierre Blanchard, 1793

The first true balloon flight in the United States occurred when Frenchman Jean-Pierre Blanchard ascended from the yard of the Washington Prison in Philadelphia on January 9, 1793. President-elect George Washington, the French ambassador and many onlookers watched Blanchard ascend to about 5,800 feet. He drifted to a landing near Woodbury in Gloucester County, New Jersey. It was Blanchard's forty-fifth balloon ascension.

Steerable Balloon—Solomon Andrews, 1863

Dr. Solomon Andrews (1806–72) from Perth Amboy was a prolific inventor. When he created a lock that that he claimed no one could open, he invited many locksmiths and bankers to test his lock in public. Indeed, no one could open it. In fact, the lock was so good that the U.S. Post Office has been using it since 1842.

After this first success, Solomon decided to improve the hot-air balloon. Since 1793, when the Montgolfier brothers first sent a balloon aloft in France, people had been enjoying hot-air balloon rides. The problem, Solomon knew, was that these craft could not be steered; they would go wherever the wind took them. And how could riders get home?

Using Perth Amboy's old barracks from the French and Indian War, Solomon set up an invention shop. His Aereon was composed of three eighty-foot, cigar-shaped balloons with a rudder for steering and a gondola for the passengers. To ascend, Andrews jettisoned (threw overboard) sand ballast; to descend, he released hydrogen lifting gas. Dr. Andrews wrote to President

Lincoln to offer the craft for use in the Civil War. About a year later, he was told that the government had little interest in the Aereon.

Solomon Andrews was a prolific Perth Amboy inventor. *Courtesy of John Kerry Dyke.*

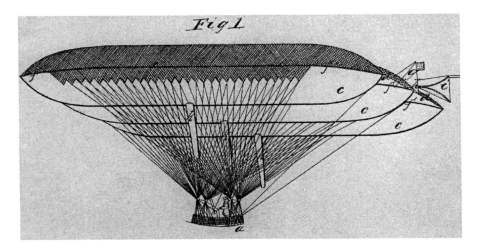

The Aereon. *Courtesy of John Kerry Dyke.*

Dr. Andrews's second version, Aereon #2, had one lemon-shaped balloon, pointed sharply at the ends. In this craft, Andrews controlled lift and descent through the use of pulleys that compressed the gas or let it expand. He flew it twice over New York City, but then his company failed. He never flew the Aereon again.

Dr. Andrews also invented a sewing machine, a barrel-making machine, fumigators, a kitchen range, a gas lamp, a nicotine-filtering pipe and a lock re-keyer.

Dr. Solomon Andrews was inducted into the New Jersey Inventors Hall of Fame in 1992.

Hovercraft—Charles J. Fletcher, 1940s

While serving as a pilot in the U.S. Navy in Norfolk, Virginia, Charles J. Fletcher sketched the design for a vehicle that could rise above the water or land approximately ten inches to two feet, depending on available horsepower. The vehicle would generate an airflow trapped against a uniform surface such as the ground or water, freeing it from the surface and eliminating friction. Aircraft control techniques and the release of air created positive control and movement. What Fletcher called the "Glidemobile" is

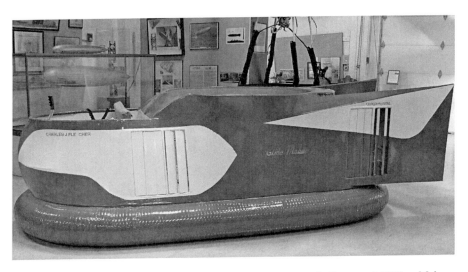

Charles J. Fletcher designed the GlideMobile, the first hovercraft. *Courtesy of William Maloney and the New Jersey Aviation Hall of Fame and Museum.*

known today as the hovercraft, which has proven to be a major advance in military land-assault vehicles and modern inter-waterway travel. Today, hovercraft are manufactured in the United States by Bell Aerosystems. They cost between $800,000 and $1.5 million each. During World War II, the military kept the idea secret, so few people knew that Fletcher had invented the hovercraft. Recently, however, British Hovercraft Ltd. claimed to have invented the craft before Fletcher, and the company sued the United States for royalties of $104 million.

Attorneys for the U.S. Department of Justice found a 1960 edition of *Design News* that featured an article on Fletcher's hovercraft. Fletcher located his records on the project, which included sixteen-millimeter films of the Glidemobile, documentation regarding his conceptual drawings, subsequent work, model flight trials and various news articles, all of which easily proved that he had created the aircraft before Hovercraft Ltd.

Fletcher earned a bachelor's degree in aeronautical engineering from the Academy of Aeronautics at New York University in 1950. He holds seventeen aeronautical patents on vertical lift and rocket engines plus five additional patents for industrial products.

Charles J. Fletcher was inducted into the New Jersey Inventors Hall of Fame in 1993.

Robot Navigator—William L. Maxson, 1945

A 1921 graduate of the U.S Naval Academy, William L. Maxson (1889–1947) resigned his commission in 1935. He established the W.L. Maxson Co. in New York City while living in West Orange. He is personally credited with nine inventions. Among these are a multiplying machine, toy building blocks and various mathematical apparatuses. One apparatus, a "robot navigator," was used by airplane navigators to compute positions in flight. Howard Hughes used the robot navigator on his famous flight around the world in 1938. During World War II, W.L. Maxson Co. developed and manufactured several important devices, including the robot navigator and the mounting system for multiple anti-aircraft guns. The men and women of the W.L. Maxson Co. were presented with the Army-Navy "E" Award for outstanding war production in September 1944.

William L. Maxson was inducted into the New Jersey Inventors Hall of Fame in 1992.

Dehmel Flight Trainer/Simulator—Richard Dehmel, 1950

Although Ed Link of Binghamton, New York, designed very early flight simulators, Dr. Richard Dehmel (1904–92) was the first to solve the equations of flight and have the controls and instruments of the trainer respond as an accurate equivalent of a real airplane.

Dr. Dehmel, working at the Curtiss-Wright Corporation in Caldwell, New Jersey, was granted his patent on January 10, 1950. The trainer/simulator greatly reduced the cost, time and risk involved in training aircraft crews. It also allowed a much higher level of training in "extraordinary situations." For example, Pan American World Airways trained 125 flight crews, plus 46 British Overseas Airways and 85 military transport crews during thirteen thousand hours of simulator time. The simulator enabled Pan Am to reduce crew training costs by 60 percent and in-flight training time from twenty-one to eight hours per crew.

Richard Dehmel was inducted into the New Jersey Inventors Hall of Fame in 1996.

Richard Dehmel invented the flight trainer/simulator, the first to solve the equations of flight and have the controls and instruments of the trainer respond as an accurate equivalent of a real airplane. *Courtesy of William Maloney and the New Jersey Aviation Hall of Fame and Museum.*

The interior of the Dehmel flight trainer. *Courtesy of William Maloney and the New Jersey Aviation Hall of Fame and Museum.*

MARINE NAVIGATION

Steam-Powered Boat—John Stevens, 1808

Colonel John Stevens III (1749–1838) of Hoboken completed his one-hundred-foot-long steamboat, *Phoenix*, in 1808. One year before, inventor Robert Fulton had built and sailed his *Clermont* along the length of the Hudson River, making it the first full-sized steamboat to operate on a regular route. Fulton's financial backer, Robert Livingston, got New York State to pass a law giving the Fulton-Livingston company exclusive rights to operate on New York waters. As a result, when Stevens's *Phoenix* was completed, it could not operate on the Hudson River. The Stevens family decided to run its passenger and freight service on the Delaware River, between Trenton

John Stevens of Hoboken invented the steamboat *Phoenix* in 1808. *Courtesy of Library Special Collections, Stevens Institute of Technology, Samuel C. Williams Library, Hoboken, New Jersey.*

and Philadelphia. To get there, the *Phoenix* had to sail on the Atlantic Ocean, making it the first steamboat in the world to sail upon the ocean.

John Stevens was inducted into the New Jersey Inventors Hall of Fame in 1989.

First Steam Ferry Service—John Stevens, 1811

In 1811, Colonel Stevens bought a license to operate a commercial ferry between Manhattan and New Jersey. While building the steamboat *Juliana* (named for his daughter), he operated a horse-powered ferry. When the *Juliana* began operations between Hoboken and New York, it became the first regularly operated steam-powered ferry service in the world.

Because of the law that Livingston had backed, Colonel Stevens was forced to stop his trans-Hudson operation and run his vessel on the Connecticut River. After Livingston's death, the law changed, and the Stevens family owned and operated the Hoboken Ferry Company, carrying commuters between New Jersey and New York City.

More Steamboat Improvements—Robert and Edwin Stevens

Robert and Edwin Stevens, sons of Colonel John Stevens, were inventors in their own right. Together, they created new and better steam engines. They also designed and constructed the first railroad tracks and the first ironclad ship.

Robert improved his father's steamboat by inventing the first concave waterlines on a steamboat (1808); the first iron rods for projecting guard beams on steamboats (1815); the first skeleton walking beams for ferries (1822); the spring pile ferry slip (1822), which yields and springs back when nudged by the boat; the placement of boilers on guards outside the paddle wheels of ferries (1822); the truss for stiffening ferry boats longitudinally (1827); spring steel bearings for paddle-wheel shafts (1828); and improved packing for pistons (1840). He was also the first to successfully burn anthracite coal in a cupola furnace (1818), the first to successfully burn anthracite coal to fire a steamboat's boilers and the first to protect the man at the wheel by enclosing him in a pilothouse.

Stevens later built the *Juliana*, which became the first regularly operated steam-powered ferry service in the world. *Courtesy of Library Special Collections, Stevens Institute of Technology, Samuel C. Williams Library, Hoboken, New Jersey.*

Edwin Stevens, Robert's younger brother, designed the "Stevens plow," a cast-iron plow with a moldboard ingeniously curved to leave no dirt sticking to it and an easily removable heelpiece. For New York City, he designed a "two-horse dump wagon" with removable sides, which the city used to haul refuse. He also developed the "closed fireroom" system of forced draft that greatly increases an engine's efficiency.

Modern Submarine—John Holland, 1878

Often called the "Father of the Modern Submarine," New Jersey inventor John Philip Holland (1841–1914) successfully launched his *Holland Boat No. 1* into the Passaic River at Paterson on May 22, 1878.

Born in Ireland, Holland came to the United States in 1873. His first design, a submersible boat powered by foot pedals, was rejected by the U.S. Navy. Much to Holland's dismay, however, the navy released his plans without his permission.

In the first trial of *Holland Boat No. 1*, the vessel descended to a depth of twelve feet and returned safely to the surface. Holland kept the submarine at the bottom for an hour during the second trial, and having succeeded, he stripped the boat of usable equipment and scuttled it in the Passaic River.

After two years of work, *Holland II* (also known as *Fenian Ram*) was launched in 1881. At thirty-one feet long, this second vessel carried a pneumatic gun in the bow and could carry a crew of three: a commander, an engineer and a gunner. The submarine was tested as far south as the Narrows (between Brooklyn and Staten Island), submerging as deep as fifty feet in the ocean.

The next vessel, *Holland III*, or the *Zalinski Boat* (named for its backer, Lieutenant Edmund Zalinski), had an accident during the launching. It was repaired and carried out several sea trials but was eventually sold.

Over the next few years, Holland designed several other submarines, winning two more navy competitions. In 1887, while living in Newark, Holland entered a U.S. Navy competition to design a submarine torpedo boat. The requirements were that the vessel had to reach a speed of fifteen knots on the surface and eight knots submerged and obtain an underwater endurance of 150 feet for two hours. Holland's design won, but the shipyard he had contracted with reneged. The navy recommended another competition, which Holland also won. Unfortunately, with a new presidential administration, the navy diverted the money to complete surface ships.

Opposite, top: Professor John Holland looks out from the hatch of the *Holland VI* submarine. *Library of Congress*.

Opposite, bottom: The *Holland VI* en route to Washington, D.C., via the Delaware and Raritan Canal. *Courtesy of the Canal Society of New Jersey*.

In 1892, Holland designed the *Holland IV*. With a new administration in the White House, Congress appropriated $200,000 to reopen the submarine competition. With associates Charles Morris and Elihu Frost, Holland formed the John P. Holland Torpedo Boat Company, agreeing to assign his earlier patents and all rights to future inventions to the company.

The 1893 competition attracted eight contenders, including Simon Lake of Pleasantville, New Jersey. Again, Holland's design was chosen, but it wasn't until March 1895 that the secretary of the navy finally awarded to Holland the construction contract for $200,000. The construction of the new submarine began in Baltimore, at the same dock at which Simon Lake's *Argonaut I* was being built. With the navy demanding changes to the design, the *Plunger*, as it was now called, ran into problems. After its 1897 launch, Holland learned that the unshielded boiler made the ship too hot. Although the navy authorized the purchase of two more subs like the *Plunger*, Holland knew that this design was a dead end. He convinced his associates at the Holland Torpedo Boat Company to build a new submarine, the *Holland VI*, as a private venture, with no navy requirements. In late 1896, while the *Plunger* was still being built in Baltimore, the keel of *Holland VI* was laid down at the Crescent Shipyard in Elizabethport, New Jersey. Work went quickly, and the boat was launched on May 17, 1897. However, when a careless workman left a valve open, salt water soaked the electrical system. The boat was finally ready for sea trials in February 1898, diving and surfacing successfully off Staten Island on St. Patrick's Day. Finally, in November 1899, *Holland VI* performed flawlessly before the Naval Board of Inspection and was purchased by the U.S. Navy for $165,000. It cruised to Washington via the Delaware and Raritan Canal in New Jersey.

John Holland was inducted into the New Jersey Inventors Hall of Fame in 2007.

Submarine—Simon Lake, 1894

Simon Lake (1866–1945) came from a family of inventors. His father, uncles and brothers were full of new ideas and creations. They were among the founders of both Atlantic City and Ocean City, New Jersey.

Born in Pleasantville, Simon joined his father's foundry and machine shop after finishing school. While working at the shore, he invented a steering gear and dredge to improve the operation of fishing and oyster-harvesting vessels.

Inspired by Jules Verne's *Twenty Thousand Leagues Under the Sea*, Lake submitted his first plans for a submarine to the U.S. Navy in 1892. In 1894, he built the *Argonaut Jr.* using pitch pine as an inexpensive way to demonstrate his principles of submergence. When submerged in shallow water, divers could open a door to exit and reenter and to retrieve articles while maintaining a pressurized compartment. His vessels had the unique feature of wheels, which kept them from getting stuck on the seafloor and provided mobility through the use of hand cranks.

Lake demonstrated his submarine at Atlantic Highlands, New Jersey, amazing onlookers and inspiring support for the Lake Submarine Company, established in 1895. In 1897 in Baltimore, he built a steel vessel, the *Argonaut I*. In 1898, with a four-man crew, the *Argonaut I* began a two-thousand-mile cruise in Chesapeake Bay and along the Atlantic coast. They traveled both on the surface and under it, retrieving fish, clams and oysters through the dive door.

Lake developed the modern submarine with an "even-keel" type submarine. This differed greatly from the *Holland* sub, which was the "sinking" type or "diving" type submersible. Simon Lake continued to design and build innovative submarines, which his company sold to both the Russian and U.S. governments. Lake's submarine career spanned over fifty years, from the 1880s until his passing in 1945.

Simon Lake was inducted into the New Jersey Inventors Hall of Fame in 2000.

RAILROADS

Steam Locomotive—John, Robert and Edwin Stevens, 1826

Twenty years after inventing the steamboat, John Stevens and his sons designed, built and tested a railroad locomotive (with a multi-tubular boiler). Imagine building a circular railroad track in your yard! That's what John did to test the engine at his estate in Hoboken. The locomotive carried six people at twelve miles per hour and was the first engine of its kind to run in the United States.

First Railroad to Cross New Jersey—Robert and Edwin Stevens and Robert F. Stockton, 1830–31

In 1830, the Stevens brothers and Robert F. Stockton received a charter from the state of New Jersey to operate two transportation companies. The Stevens brothers were granted the Camden & Amboy Railroad and Transportation Company, while Stockton was granted the Delaware & Raritan Canal Company. Both would travel the same approximate route, from Bordentown on the Delaware River to the Raritan River. The railroad would terminate in Perth Amboy, while the canal ended in New Brunswick.

Robert Stevens then sailed for England to buy a locomotive and railroad track. While en route, he conceived the flanged, T-rail type of track, on

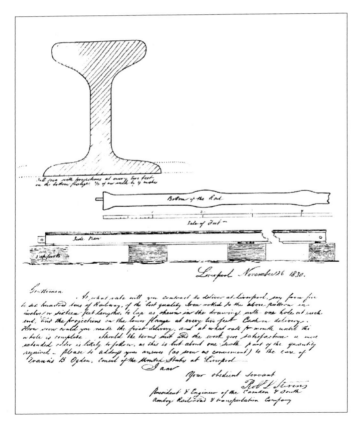

Robert Stevens is considered to be the inventor of the T-rail shape for railroad tracks. This shape is now used by railways of every nation. *Courtesy of Library Special Collections, Stevens Institute of Technology, Samuel C. Williams Library, Hoboken, New Jersey.*

which trains travel today. Used by railways of every nation, it replaced the cast-iron edge rails without flanges that had been introduced in England in 1789. He created the first all-iron rail construction for the Camden & Amboy Railroad of New Jersey. (Before 1831, the rails of all previous American railroads were strap-iron rails made of wood with a metal strap applied to the wood.) By 1834, the *John Bull* locomotive, the first efficient passenger locomotive in the United States, was traveling between Perth Amboy and Camden.

Seeing that European railroads had passenger cars that were entered from the side with no connection between compartments, Edwin Stevens introduced the vestibule car, which had a center aisle, a door at each end and a vestibule for getting on or off and for passing to the next car. In his will, Edwin provided for the establishment of Stevens Institute of Technology in Hoboken.

AUTOMOBILES

Track-Laying Vehicle—Jesse Lake, 1860s

The inventor of the track-laying vehicle, Jesse Lake, helped build the first highway between the mainland and Atlantic City in the 1860s. These vehicles, similar to today's bulldozers, tanks and construction machines, carried sand, gravel and other materials across the soft meadows and streams.

Three-Wheeled Automobile—C. W. Kelsey, 1898

In 1898, while attending Haverford College in Haverford, Pennsylvania, C.W. Kelsey of Short Hills, New Jersey, and Sheldon Tilney built a three-wheeled automobile that they called the Auto-Tri. Kelsey, whose career spanned twenty years, followed the Auto-Tri with the Motorette, a small, self-propelled vehicle between the automobile and the motorcycle. It had the low cost of a motorcycle and the road ability of a larger car. In 1921,

Kelsey formed the Kelsey Motor Co. in Newark, New Jersey, where he built the Kelsey sedan coupe, Roadster and touring car. The company produced the four-passenger sedan coupe in 1922 and 1923. He called this automobile the Standard Gear Car. The 1923 Kelsey five-passenger touring car had a dark-green finish and a leather interior, a one-man never-leak top, a Stewart-Warner speedometer, a spare tire, a tool kit, a pump, a tire repair kit, a starting crank, a rim wrench, switch keys and curtain rods. The price of the Kelsey touring car was $1,800, the Kelsey roadster $1,750 and sedan $2,700.

Universal Joint—Clarence W. Spicer, 1903

In 1903, Clarence W. Spicer (1875–1939), a graduate of Alfred University in New York, received a patent for the universal joint, a critical component of the modern automobile. He made joints and axles at the Mack Truck plants in Plainfield and New Brunswick, New Jersey, from 1904 to 1914 and in South Plainfield from 1914 to 1930. Much of his South Plainfield factory still stands. Spicer received the patent while continuing his studies at Cornell University and began manufacturing his invention in 1904 at the Spicer Manufacturing Company in Plainfield. Before his company even opened, the joint was in demand by the automobile industry. Early cars used sprockets and chains, much like bicycles. They were unsightly and noisy at high speeds. They were also difficult to lubricate, and they broke frequently, no doubt due to stones and rocks, as there were only 144 miles of paved roads in the country at that time. Spicer's invention replaced the chain method of transferring power from the engine to the wheels. The universal joint attached the engine and rear axle to a propeller shaft, reducing noise. It also protected the vehicle against dust and dirt and was easy to lubricate.

In need of capital, Spicer went to an investment banker in New York, where he met Charles Dana. Seeing the promise of the company, Dana reorganized and refinanced Spicer Manufacturing, allowing Spicer to concentrate on product improvement and the invention of new products. As president, Dana led the company to become a multifaceted automotive supplier. Now, more than a century later, Spicer's company is part of the Dana Corporation in Toledo, Ohio.

Clarence Spicer is a member of the Automotive Hall of Fame.

First Traffic Circle, Pennsauken, Camden County, 1925

When the Airport Circle was built in 1925, it allowed traffic to circulate through multiple intersections without the need for a traffic light. Circles, or rotaries, worked well until after World War II, when increased traffic in the suburbs and the refusal by some drivers to give up the right-of-way in the circles made driving through them dangerous. The New Jersey Department of Transportation has been gradually modifying or eliminating traffic circles to make driving safer.

Airport Circle, the country's first traffic circle, was built in 1925. It allowed traffic to circulate through multiple intersections without the need for a traffic light. *New Jersey State Archives.*

First Cloverleaf Interchange, New Jersey Department of Transportation, 1929

A cloverleaf interchange is used mainly on heavily traveled roads. It allows traffic to turn left by using loop roads or ramps that pass over or under the main highway. The first cloverleaf interchange in the United States was built in Woodbridge Township, New Jersey, in 1929. It connected Routes 25 and 4, now known as U.S. Highway 1/9 and Route 35, respectively. It has since been replaced with a partial cloverleaf interchange. The original was modeled after a plan from Buenos Aires, Argentina.

The first cloverleaf interchange was built in Woodbridge, New Jersey. Now common on interstate highways, these interchanges allow traffic to turn left by using loop roads or ramps that pass over or under the main highway. *New Jersey State Archives.*

The Jersey Barrier—Stevens Institute of Technology, 1959

A Jersey barrier is a modular concrete barrier employed to separate lanes of traffic. These structures were designed to lessen vehicle damage if accidentally hit while still preventing the crossover type of head-on collision. Jersey barriers are also used to reroute traffic and protect pedestrians during highway construction. Known in the western United States as K-rails, the barriers are now also made of plastic. According to the Hunterdon County Historical Society, frequent accidents on the steep road over Jugtown Mountain in Hunterdon County led to the first installation of the Jersey barrier. These concrete dividers are now seen at construction sites throughout the world. More recently, they have been used to defend against terrorists.

Also called the New Jersey wall, the barrier was developed at the Stevens Institute of Technology in the 1950s (introduced in its current form in 1959), under the direction of the New Jersey State Highway Department to divide multiple lanes on a highway. A typical Jersey barrier stands thirty-two inches tall and is made of steel-reinforced concrete or plastic. Many are constructed with the embedded steel reinforcement protruding from each end, allowing them to be incorporated into permanent emplacements when linked to one another.

Jackknife Protection and Ignition Shield—Joseph J. Mascuch

A resident of Short Hills, New Jersey, Joseph Mascuch (pronounced ma-SHOO) was awarded 165 patents, many for devices to improve various aspects of transportation. He is best known for the development of the ignition shield, which allows the clear reception and transmission of radio signals when the electrical system of an airplane or automobile engine is running. Mascuch developed an anti-jackknife hitch that prevents tractor-trailers from jackknifing when braking, as well as a flexible metal hose that allows the use of safe fuel lines in aircraft, a device first used in Wiley Post's airplane, the *Winnie May*, in 1934. Among his other devices are rustproof automobile bumpers, a propeller blade, a marine bulkhead door, an antenna structure, standardized helicopter hoists, the mechanism for carrying and releasing bombs and rockets, the instant reading thermocouple thermometer and gear-driven clamps.

Joseph J. Mascuch was inducted into the New Jersey Inventors Hall of Fame in 1995.

BICYCLES

Mount Holly and Smithville Bicycle Railway— Arthur E. Hotchkiss, 1892–98

Arthur Hotchkiss designed a monorail transportation system to connect the town of Mount Holly with the village of Smithville in Burlington County. He arranged for the Smith Machine Company to build the prototype of his system, which was tested on a two-hundred-foot track at the factory. Hotchkiss then proceeded to construct a 1.8-mile bikeway, which crossed a creek ten times to afford the users scenic views. The track was built with an upside-down T-rail set along the top of a fence. Specially built bicycles, designed to straddle the monorail, ran along the track. A series of small guide wheels at the bottom kept the frame of the bicycle steady and erect.

By late 1892, the bicycles and track were completed, and Hotchkiss announced that his railroad was open for business "every night and day except Sunday." During the week of the county fair, over five thousand people rode the bicycle railroad. The *Mount Holly Herald* noted that "the cash receipts have been sufficient to pay one year's interest on the bonds." The paper further reported:

> *Every night, there is a crowd of people at the depot waiting their turn for a ride, and the machines are kept busy until eleven o'clock at night. To say that Prof. Hotchkiss is delighted at the success of his invention does not half express it. The roadway is illuminated at night, and each machine carries a light to avoid collisions. Derailment of a machine is impossible, and the road is absolutely safe.*

Hotchkiss had the privilege of demonstrating his railroad at the World's Columbian Exposition in 1893. Others acquired the rights for the system in Atlantic City, Ocean City and Gloucester, New Jersey. Not just an entertainment, the railroad carried workers from Mount Holly to the H.B. Smith Machine Company in Smithville. A monthly commuter ticket cost $2.00, and at the end of ten months, gross receipts totaled $1,893.81, with a net income of $800.00. Gradually, however, the railroad fell into disrepair, and by 1898, with a slowdown at the factory, Hotchkiss decided to give it up.

CHAPTER 8
WHAT'S UP, DOC?: MEDICAL ADVANCES

THE BAND-AID—EARLE DICKSON, 1920

"I am stuck on Band-Aid brand, 'cause Band-Aid's stuck on me." This jingle from the Johnson & Johnson commercial would not have been possible without the determination of J&J employee Earle Dickson, a cotton buyer for the company.

Back in 1920, Earle and his wife, Josephine, were living in New Brunswick, New Jersey. Josephine enjoyed making dinner for her husband, but unfortunately, she often got cuts or burns while she was preparing and cooking the food. Since she had no easy way of bandaging her own cuts, Earle had to cut pieces of adhesive tape and pieces of cotton gauze. He then covered the cut with the gauze and covered the gauze with the tape, making a bandage for each wound. This happened day after day, and day after day, Josephine needed more bandages. What could they do?

Finally, after several weeks of kitchen accidents, Earle hit upon an idea. (Luckily for Johnson & Johnson, his idea was not to go out and hire a cook.) He prepared some ready-made bandages by placing a strip of cotton gauze down the center of an adhesive strip and covering the strip with crinoline (a coarse, stiff fabric of cotton or horsehair used especially to line and stiffen hats and garments). Now all Josephine had to do was cut off a length of the strip and wrap it over her cut.

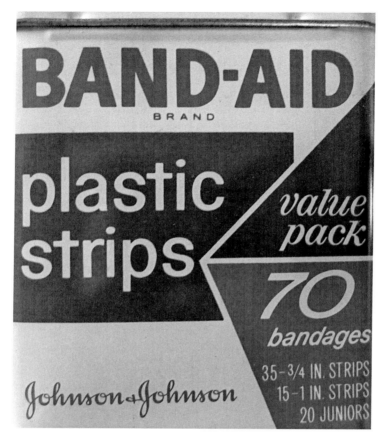

Johnson & Johnson began making Band-Aid brand adhesive bandages in the 1920s. *Courtesy of Robert H. Barth.*

Earle told his boss at J&J about his new invention, and soon the first adhesive bandages were being produced and sold under the Band-Aid trademark. They were made by hand and were not very popular at first. Why? Maybe because they were three inches wide and eighteen inches long! People had to cut them apart, just as Mrs. Dickson had done. Only $3,000 worth of Band-Aids were sold that year. To boost sales, Johnson & Johnson decided to give free Band-Aids to Boy Scout troops as a publicity stunt. By 1924, Band-Aids were machine made. Earle was eventually rewarded with a position as vice-president at the company, where he stayed until his retirement. As for Josephine, we don't know whether she ever learned to cook without accidents. But we do know she had plenty of Band-Aid brand adhesive bandages available just in case.

New Jersey Low Contact Stress (LCS) Total Knee Replacement—Michael J. Pappas and Frederick Buechel, 1970s—'80s

Michael J. Pappas designed a knee implant known as the New Jersey Knee, a replacement for damaged knees. Working with Frederick Buechel, Pappas invented the New Jersey Low Contact Stress Total Knee Replacement, which has helped thousands of patients walk without pain after an injury or illness. Pappas was teaching biomechanics at the University of Medicine and Dentistry of New Jersey in 1974 when he asked Buechel to help him build a better knee implant. The two men have also developed replacement systems for the ankle, hip and shoulder.

Dr. Pappas received his PhD in 1970 from Rutgers University in mechanical engineering, specializing in computer-aided structural design. He has been a professor of mechanical engineering at the New Jersey Institute of Technology for more than twenty-five years.

Michael Pappas was inducted into the New Jersey Inventors Hall of Fame in 1998.

Streptomycin—Albert I. Schatz and Selman Waksman, 1943

In 1943, Albert I. Schatz (1920–2005), a twenty-three-year-old graduate student at Rutgers University, and his teacher Selman Waksman co-discovered streptomycin, a powerful antibiotic hailed as a miracle drug. It was the first effective treatment against tuberculosis, a disease that had killed more than one billion people over the previous two centuries.

Tuberculosis (TB) is an infectious disease that usually infects the lungs but can attack almost any part of the body. It is contagious, spreading through the air through sneezing or coughing. Symptoms include a cough that is worse in the morning, chest pain, breathlessness, night

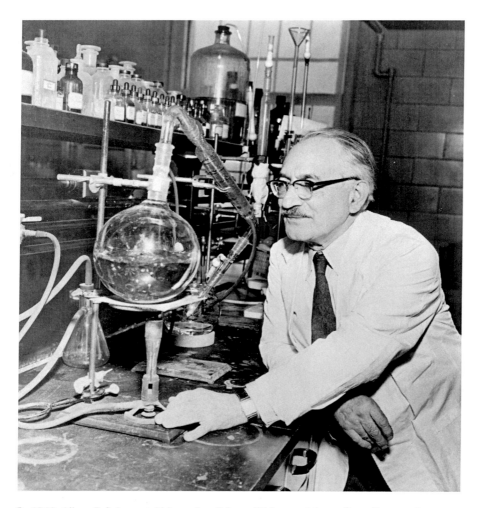

In 1943, Albert I. Schatz and his teacher Selman Waksman (pictured) co-discovered streptomycin, a powerful antibiotic hailed as a miracle drug. *Library of Congress.*

sweats and signs of pneumonia. In advanced cases of the disease, there may be extreme weight loss. TB was once the leading cause of death in the United States.

Streptomycin was also effective against other organisms, including those that cause typhoid and plague. This discovery was important for another reason: at that time, some organisms did not respond to the penicillin and sulfa drugs that were then available.

At the time of his discovery, Dr. Schatz was working in the Rutgers laboratory of Dr. Waksman, an eminent microbiologist whose laboratory was developing new antibiotics. Dr. Schatz volunteered to search for an antibiotic that could be used to fight tuberculosis and worked alone in the basement to reduce the risk of infection to his colleagues. After three and a half months, he isolated the antibiotic that became known as streptomycin and wrote the first paper announcing the discovery.

Selman Waksman was awarded the Nobel Prize in Physiology and Medicine in 1952. He was inducted into the New Jersey Inventors Hall of Fame in 1989.

Albert Schatz was awarded the Nobel Prize in Physiology and Medicine in 1952. He was inducted into the New Jersey Inventors Hall of Fame in 2005.

Microorganism "Shaking" Machine—David Freedman

Dr. Selman Waksman and others from the Rutgers microbiology department needed someone to repair their "shaking" machine, a device that constantly moved bacterial cultures so they would not congeal. They hired New Brunswick inventor David Freedman, who owned New Brunswick Scientific Company. Freedman created a better machine, which soon became the mainstay of his firm's product line. Freedman, who had opened his business in 1946, saw it grow into a global biotechnology equipment maker. His machines helped grow, detect and store microorganisms used in research. He secured twenty patents for equipment that was used in cancer research and pollution abatement.

New Antibiotics—Selman Waksman, 1940–48

Born in Russia, Selman Waksman (1888–1973) entered Rutgers College in 1911, having won a state scholarship the previous spring. He received his BS degree in agriculture from Rutgers in 1915 and was appointed research assistant in soil bacteriology at the New Jersey Agricultural Experiment Station. Waksman was allowed to continue graduate work at Rutgers, obtaining his MS degree in 1916.

Together with his students and associates, Dr. Waksman isolated a number of new antibiotics, including actinomycin, clavacin, streptothricin, streptomycin, grisein, neomycin, fradicin, candicidin, candidin and others. Two of these, streptomycin and neomycin, have been used in the treatment of numerous infectious diseases of men, animals and plants. The patent on streptomycin has been listed as one of the ten patents that shaped the world.

DNA Test for Tuberculosis— Dr. David Alland, 2010

Dr. David Alland invented a new test to determine whether a patient has tuberculosis (TB), a contagious bacterial infection that usually attacks the lungs but may spread to other organs such as the kidney, spine and brain. If not treated properly, TB can be fatal. Dr. Alland's new test shortens the time it takes to get the results of the test from three months to less than two hours. Dr. Alland, the chief of the Division of Infectious Disease at the University of Medicine and Dentistry in Newark, was pleased that the new test had been endorsed by the World Health Organization.

PREDNISONE AND PREDNISOLONE— ARTHUR NOBILE, 1950–51

When Arthur Nobile patented prednisone and prednisolone, he created drugs that would go on to alleviate much suffering and save many lives. A researcher who mainly worked alone, Nobile (1920–2004) worked for Schering Corporation (later Schering-Plough, now part of Merck) and succeeded in using bacteria to oxidize cortisone into prednisone and hydrocortisone into prednisolone. These extracts were more than four times more effective than natural cortisones against arthritis in mice. Nobile had proven that natural compounds can be improved upon through molecular biology. His work led to the development of many life-saving drugs. Modifications of the prednisolone molecule, for example, have resulted in compounds to treat asthma, psoriasis, ulcerative colitis, cerebral edema caused by cancer and skin disorders.

Arthur Nobile was inducted into the New Jersey Inventors Hall of Fame in 2000.

APGAR SCALE—VIRGINIA APGAR, 1952

A resident of Westfield, Dr. Virginia Apgar was a pediatric anesthesiologist who developed a measurement scale that allows doctors and nurses to quickly evaluate the health of newborns. The scale has greatly reduced infant mortality around the world. The Apgar scale evaluates the baby's complexion (skin color), pulse rate, reflex irritability (response when stimulated, such as a cry or grimace), muscle tone and breathing. The infant's response to each of these areas produces a "bacronym" using the letters in the doctor's name:

Appearance (skin color)
Pulse (rate)
Grimace (response to stimulation)
Activity (muscle tone)
Respiration (breathing)

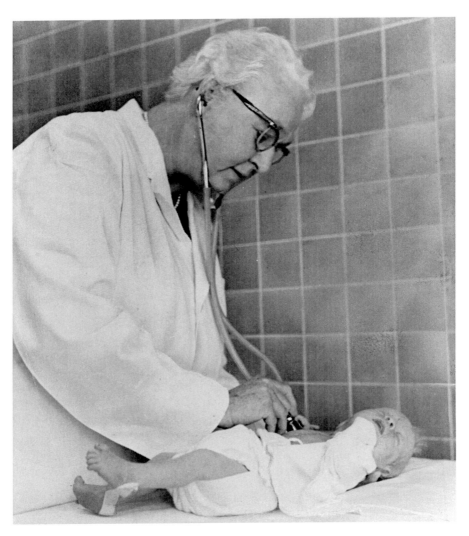

Dr. Virginia Apgar developed a measurement scale that allows doctors and nurses to quickly evaluate the health of newborns. The scale has greatly reduced infant mortality around the world. *World Journal Tribune* photo by Al Ravenna. *Library of Congress.*

GENTAMICIN—MARVIN WEINSTEIN, 1963

Gentamicin is one of the most widely used and important antibiotics available today. It was discovered by Marvin Weinstein while he was working at Schering-Plough (now part of Merck). The holder of twenty-eight patents, Weinstein is also the co-inventor of the process for extracting interferon from bacteria. Interferons are naturally occurring proteins that are made and secreted by cells of the immune system (for example, white blood cells, natural killer cells, fibroblasts and epithelial cells). They boost the immune system response and reduce the growth of cancer cells by regulating the action of several genes that control the secretion of numerous cellular proteins that affect growth.

Marvin Weinstein was inducted into the New Jersey Inventors Hall of Fame in 1990.

TETRACYCLINE—HERMAN SOKOL, 1955

Dr. Herman Sokol is a co-discoverer of tetracycline and a pioneer in the production of antibiotics. He was the president of the Bristol-Myers Company from 1976 to 1981. Montclair State University has established the Margaret and Herman Sokol Institute for Pharmaceutical Life Sciences in his honor. Tetracycline is used for treating several types of infections caused by susceptible bacteria. Some examples include infections of the respiratory tract, urinary tract and skin. It is also prescribed for nongonococcal urethritis, Rocky Mountain spotted fever, typhus, chancroid, cholera, brucellosis, anthrax, syphilis and acne. Tetracycline is used in combination with other medications to treat the bacteria associated with ulcers and inflammation of the stomach and duodenum.

Herman Sokol was inducted into the New Jersey Inventors Hall of Fame in 2008.

TheraGrip—John J. Frins

People who have multiple sclerosis or have had injuries to a hand or lower arm must do exercises to strengthen those muscles. John J. Frins, a South Orange resident, invented the TheraGrip hand strengthener, which is capable of exercising all five fingers. It is the only such device that involves the thumb, and this unique factor was the basis of his patent. The TheraGrip, manufactured by Baxter Rubber Company, allows disabled individuals to exercise and continue to lead more normal and productive lives.

John J. Frins was inducted into the New Jersey Inventors Hall of Fame in 1994.

ZoltanCAP —Bart Zoltan

Working at the American Cyanamid Company in Bound Brook, New Jersey, Bart Zoltan invented the Compliance Aid for Pharmaceuticals, known as the ZoltanCAP. Built into standard pill bottles, this aid shows the time and date of the last time the container was open, allowing patients to keep track of their prescribed medications.

Bart Zoltan was inducted into the New Jersey Inventors Hall of Fame in 1990.

Bart Zoltan invented the Compliance Aid for Pharmaceuticals, known as the ZoltanCAP. *Courtesy of Bart Zoltan.*

CHLORINE PUMP—CHARLES FREDERICK WALLACE, 1913

Until the early 1900s, people often contracted diseases from drinking or swimming in contaminated or polluted water. Scientists knew that adding chlorine to the water would kill most of the harmful bacteria. But how could they add it to large reservoirs?

To solve this problem, Charles Frederick Wallace invented the chlorinator, which provided the first practical means to sterilize drinking water by injecting chlorine gas into it. The pump was first used at the Boonton, New Jersey reservoir from which drinking water was piped to Jersey City. At the time, pollution from a small stream was threatening the water supply. Martin F. Tiernan, Wallace's partner, convinced Jersey City's water department that the chlorinator could solve the city's pollution problems for only $150. The device was installed in a blacksmith's shop near the reservoir, but a gas leak turned the blacksmith's tools green, so he threw the device into the reservoir. After fishing the pump out of the water, Wallace hooked it up again, and it worked properly.

Wallace's first invention was so successful that within a few years, the Wallace & Tiernan device was being used to purify half the world's drinking-water supply.

Charles Frederick Wallace was inducted into the New Jersey Inventors Hall of Fame in 1996.

TINNITUS TREATMENT—ALFONSO DIMINO

Tinnitus is a hearing condition in which a person hears a constant ringing, buzzing or other noise in the ear. Alfonso DiMino (1920–2001) invented the Aurex-3, a non-invasive electronic therapy system for treating these disorders. It consists of a tabletop control unit and a hand-held applicator. The probe tip of the applicator is placed behind the patient's ear near the mastoid bone. The unit, which is not worn, can be used three to four times per day for three to five minutes each time. After earning a doctorate in chemistry from the University of Palermo, Italy, Dr. DiMino immigrated to the United States in 1950, where he worked as a handyman for a carbon-paper manufacturer in Brooklyn at $1.10 an hour. He suggested an idea to the company president for a more efficient machine to make carbon paper. Within a year, he was appointed research director for the company and earning $30,000.00 annually (the average salary in the 1950s was $2,992.00).

POWER HAWK—WILLIAM HICKERSON, 1990S

After a serious traffic or airplane accident, people often become trapped inside the crushed metal of the vehicle. To help extricate the victims from the wreckage, power spreader and cutter rescue tools (commonly referred to as the "Jaws of Life") are used by rescue and emergency responders. To improve the portability and versatility of traditional rescue tools, William Hickerson invented the Power Hawk P-16 Rescue System while he was vice-president of operations at Curtiss-Wright Flight Systems in Fairfield, New Jersey, and perhaps more importantly, the fire chief of Hardyston Township.

The Power Hawk uses Curtiss-Wright aerospace technology and twelve volts of DC battery power to deliver its enormous forces without hydraulics or the need for gasoline-powered engines. This self-contained and portable rescue system, which provides for quick-change spreader and cutter attachments and a seventy-degree pivot of its powerhead, is used by fire departments, rescue squads, police, military and other emergency responders around the world to spread, cut, crush and lift wreckage or machinery out of the way so that injured persons can be quickly removed.

Subsequent inventions by William Hickerson include the Power Pusher Ram and Auto Crib-It Automatic Vehicle Stabilization Tool. William Hickerson is currently president of Power Hawk Technologies Inc., located in Rockaway, New Jersey.

William Hickerson was inducted into the New Jersey Inventors Hall of Fame in 1999.

Opposite, top: The Power Hawk Rescue System is used by fire departments and rescue squads to spread, cut and lift machinery out of the way so that injured persons can be removed. *POWER HAWK Technologies Inc.*

Oppsote, bottom: The Power Hawk Rescue System at work. *POWER HAWK Technologies Inc.*

Communication Systems for the Handicapped— Haig Kafafian

Many people are unable to type on a standard keyboard because of an injury, birth defect or disease such as cerebral palsy. Haig Kafafian has invented communications systems that allow people with limited mobility to lead productive lives and develop their talents. A person who can control even one single part of his or her body can now communicate effectively using a keyboard with only fourteen keys. This is possible because Kafafian applied the principles of cybernetics—the science of control—to create some of the first communications devices for the handicapped.

Haig Kafafian was inducted into the New Jersey Inventors Hall of Fame in 1999.

Wiktor Stent for Heart Surgery— Dominik Wiktor, 1988

Dominik Wiktor of Cranford, New Jersey, worked as an electrical engineer for Bellcore in Morristown. In 1984, he underwent open-heart surgery to correct a problem with his aorta. Afterward, Wiktor wondered if this artery repair could have been done with less trauma. As he read more about a procedure called angioplasty (repairing a blood vessel using a catheter or the patient's own tissue), he realized that stents would be a better way to perform the surgery. Wiktor created several designs and in 1988 signed a contract with Medtronic to produce these stents. The Wiktor stent keeps a blood vessel open and prevents it from closing again. It is made of a metal called tantalum. Within a month after being inserted into the blood vessel, the stent becomes part of the artery wall.

Dominik Wiktor was inducted into the New Jersey Inventors Hall of Fame in 1996.

Mobility Chair—Michael J. Flowers, 1983

Michael J. Flowers, president of the Electric Mobility Corporation, developed the Rascal ConvertAble, a combination three-wheel electric scooter and four-wheel power wheelchair for the physically challenged. Flowers designed an add-on two-wheel front section that could be easily attached (without tools) to a two-wheel take-apart rear section. The result was the Rascal ConvertAble, offering both a sturdy outdoor three-wheel scooter and a precision four-wheel version, allowing the user to drive up to a table to work or eat without needing to swivel the chair or move anything out of the way. The combination achieved the maximum degree of mobility and eliminated the need to buy two separate units. More than five thousand Rascal ConvertAbles have been sold by Electric Mobility Corporation in various configurations. In 2008, Flowers received a patent for a foldable mobility chair.

Michael Flowers was inducted into the New Jersey Inventors Hall of Fame in 1993 and again in 2009.

Intelligent First-Aid Kit—Dave Hammond

During the Vietnam War, Dave Hammond learned that many people are not trained to deal with injuries. He was determined to create an intelligent first-aid kit that almost anyone can use, even under great stress. Using techniques found in computer graphics and filmmaking, he designed kits that are color-coded and use icons to show caregivers what to do in nearly every emergency. Each kit contains supplies and color-coded picture cards that provide easy-to-follow instructions for each specific injury—from breathing problems to bleeding. Kits are used by major organizations such as Hertz, Marriott, CBS, Disney and the U.S. Postal Service. Hammond holds a master's degree in education from George Washington University and is the author of the 1974 book *A Guide to Medical Care in Isolated Environments*.

Dave Hammond was inducted into the New Jersey Inventors Hall of Fame in 2002.

CANCER RESEARCH—GERTRUDE B. ELION

When her grandfather died from cancer, Gertrude Elion was moved to work toward a cure for the terrible disease. Entering Hunter College in 1933, she majored in chemistry. After attending graduate school at New York University (she was the only woman in her graduate chemistry class), Elion taught school and did her research work at night and on weekends, obtaining her MS in chemistry in 1941. She synthesized and co-developed two of the first successful drugs (thioguanine and mercaptopurine) for the treatment of leukemia, as well as azathioprine (Imuran), an agent to prevent the rejection of kidney transplants and to treat rheumatoid arthritis. Trudy, as she was called by her many friends, also played a major role in the development of allopurinol for the treatment of gout and of acyclovir, the first selective antiviral agent that was effective against herpes infections.

GAS PERMEABLE CONTACT LENS—NORMAN GAYLORD

The holder of three hundred domestic and foreign patents, Norman Gaylord worked to develop polymer-chemical and other products and to teach those skills to new scientists. Working with optical polymer resins, he created a scratchproof material for airplane windshields that he was later able to develop into contact lenses. Gaylord's work also played a role in the development of Saran Wrap, Teflon and the lithium battery.

CHAPTER 9

MADE IN NEW JERSEY: MANUFACTURING

WIRE ROPE CABLE—JOHN A. ROEBLING, 1841

John A. Roebling (1806–1869) came to the United States from Germany in 1831. He was soon employed to work on the Pennsylvania Main Line Canal. To transport boats over the Allegheny Mountains, the canal company had constructed a series of inclined planes. Boats were hauled on railroad track up and down the steep terrain using heavy rope, nine inches in diameter. Realizing how dangerous this was when the rope broke, Roebling invented something much stronger: flexible wire rope. He convinced the canal company to replace the large, expensive ropes with his wire cable, just over an inch in diameter.

The wire rope cable worked so well that in 1841, Roebling built a factory in Saxonburg, Pennsylvania, to make as much as the inclined planes required. He received a patent in 1842 for "new and improved mode of manufacturing wire ropes." Roebling soon found other uses for his invention. His first wire-rope suspension bridge (1844–45) was an aqueduct that carried the Main Line Canal across the Allegheny River into Pittsburgh. Receiving more patents in 1846 and 1847, he built the Delaware Aqueduct, which carried the Delaware and Hudson Canal over the Delaware River between New York and Pennsylvania. It is the oldest surviving suspension bridge in the United States (and beautifully restored by the National Park Service). Roebling

The Brooklyn Bridge was designed by John A. Roebling. *Library of Congress.*

developed a method of spinning the heavy cables at the construction site and created a simple plan to anchor them. These innovations made it much easier to build long suspension bridges. In 1848, Roebling moved his operations to Trenton, New Jersey. He had chosen this location for several reasons: it was near the Cooper Iron Works, his source of wire; it had road, canal, rail and riverboat transportation; and it was near the East Coast markets. From this base, his firm designed and built the Niagara River Gorge Bridge (1855), the Sixth Street Bridge in Pittsburgh (1859) and the Covington–Cincinnati Suspension Bridge (1856–57).

The John A. Roebling's Sons Company's last and greatest project is the Brooklyn Bridge, which spans New York's East River, connecting Manhattan and Brooklyn. When completed in 1883, the bridge was the longest suspension bridge in the world. In July 1869, soon after construction began on the Brooklyn Bridge, John Roebling died from tetanus, which he contracted when his foot was crushed in an accident on site. Almost immediately, Roebling's thirty-two-year-old son and partner, Washington A. Roebling, was named chief engineer in his place. Then, just three years later, Washington Roebling nearly died from caisson disease (the bends) after he came up too quickly from inside the underwater foundations. Washington became an invalid and was housebound for the final eleven years of the project. Fortunately, Emily, his smart and capable wife, became his assistant and communicated his ideas to the assistant engineers, bridge trustees and contractors on the project. It was she who steered the project to completion, managing political bickering, technical difficulties, contractor fraud and numerous attempts to oust her husband. Washington called her role "invaluable."

John A. Roebling was inducted into the New Jersey Inventors Hall of Fame in 2002.

FIRST U.S. SILK MILL—JOHN RYLE, 1842

John Ryle (1817–87) was born near the English silk-making city of Macclesfield. From the age of five, he worked in the silk mills, starting as a bobbin boy and eventually becoming an expert weaver. Coming to the United States in 1839, Ryle was hired to establish a silk factory in Paterson, New Jersey. He became the owner of that factory seven years later. Ryle, known as the "Father of the Silk Industry" in Paterson, was the first to successfully manufacture silk in the United States, producing the first skein in 1842.

FIRST CAST-IRON PLOW—CHARLES NEWBOLD, 1797

Charles Newbold, a blacksmith in Chesterfield, New Jersey, wanted a new method for tilling the soil. In 1797, he patented the nation's first cast-iron plow. But he was not able to sell it because neighboring farmers thought the iron plow would poison the soil. Ten years later, David Peacock received a patent for a three-piece iron plow (Newbold's was one-piece). Newbold sued Peacock for patent infringement and won $1,500.

Revolver—Samuel Colt, 1835–36

As a boy, Samuel Colt (1814–1862) liked to tinker with mechanical devices. He especially enjoyed working on his father's firearms, disassembling and reassembling them. In 1835, Samuel Colt obtained his first European patent on his revolver, and in 1836, he patented his invention in the United States. In that same year, Colt began the Patent Arms Manufacturing Company, one of Paterson's most famous industries. The Colt Gun Mill, a four-story building, was built directly below the Great Falls of the Passaic River. Here, Colt manufactured his repeating firearm, the revolver, which had mother-of-pearl handles. Between 1836 and 1841, about five thousand rifles and revolvers were made at the mill. But Colt's revolving-cartridge firearm was slow to gain acceptance, and the business closed in 1842.

Woodworking Equipment and the Star Bicycle— Hezekiah Bradley "H.B." Smith, 1865

Born in Vermont, Hezekiah Bradley Smith (1816–87) was trained as a cabinetmaker. Moving to Boston and then Lowell, Massachusetts, he designed and manufactured woodworking machinery.

Jonathan and Samuel Shreve purchased land in Burlington County in the 1830s and established a textile manufacturing facility. Named for the brothers, Shrevesville prospered until the textile depression in the 1850s.

In 1865, Smith visited the Shreves in Medford, New Jersey, and for $20,000 acquired the former textile-manufacturing town. Smith and his wife soon made this village their home, renaming it Smithville. Hezekiah ran the successful business manufacturing woodworking machinery and published the *New Jersey Mechanic*, a trade journal, while Agnes practiced medicine and sold herbal remedies.

Smith was an inventor with over forty patents, including woodworking machinery, a tenoning machine and power morticers. He incorporated his business as the H.B. Smith Machine Company. In addition to his now-mechanized woodworking business, he also built the American Star Bicycle, which had a large, back wheel under the rider's seat and a smaller front wheel for steering. Operated by a ratchet drive, the bike had a hand brake and a sprung leather seat and sold for $150. Although not a commercial success, it was a great stunt bike and an advertising tool for his company.

BIBLIOGRAPHY

Beckwith, Charles S. "Cranberry Separators." *Whitesbog Preservation Trust Newsletter* (June–August 2012).

Bellis, Mary. "History Timeline of the Battery." http://inventors.about.com/od/bstartinventions/a/History-Of-The-Battery.htm.

———. "Teflon–Roy Plunkett." http://inventors.about.com/library/inventors/blteflon.htm.

———. "Valerie Thomas." http://inventors.about.com/library/inventors/bl_Valerie_Thomas.htm.

Bolger, William C. "Elizabeth White and Historic Whitesbog." Unpublished, 1981.

Burstyn, Joan N. *Past and Promise: Lives of New Jersey Women.* Metuchen, NJ: Scarecrow Press, 1990.

Campbell, Carol Ann. "James Hillier, 91, Was a Man of Vision." *Star-Ledger,* January 20, 2007.

Casiano, Jonathan. "An Ace in the Hole." *Star-Ledger,* May 1, 2005.

Cohen, Jason. "Dripsters." *Courier-News,* July 12, 2012.

Cunningham, John T. *Newark.* Newark, NJ: New Jersey Historical Society, 1966.

————. *New Jersey: A Mirror on America.* Florham Park, NJ: Afton Publishing Co., 1978.

————. *This Is New Jersey.* New Brunswick, NJ: Rutgers University Press, 1968.

DiUlio, Nick. "Visionary Vineland." *New Jersey Monthly* (August 2011): 116.

Eschenbah, Stephen. "Home-Late Security." *Harvard Magazine,* (July–August 2004).

FeministVoices.com. "Lillian Gilbreth." http://www.feministvoices.com/lillian-gilbreth.

Frassinelli, Mike. "Ferry Service Improved, Not New." *Star-Ledger,* December 8, 2011

Gertner, Jon. *The Idea Factory: Bell Labs and the Great Age of American Innovation.* New York: Penguin Press, 2012.

Goldberg, Vicki. "It's a Leonardo? It's a Corot? Well, No, It's Chocolate Syrup." *New York Times,* September 25, 1988.

Goldman, Adam. "Some Claim Inventor Lemelson a Fraud." ABC News, August 20, 2005.

Graham, Ian. *Transportation.* New York: Bookwright Press, 1987.

Gruen, Abby. "Sun Shines Bright on Jersey Solar Panels." *Star-Ledger,* October 5, 2010.

Hansen, Susan. "Breaking the (BA) CODE." IP Law and Business Inc., (March 2004): 2–5.

Heyboer, Kelly. "Ex-Bell Labs Scientists Awarded Japan Prize for Developing UNIX." *Star-Ledger,* January 26, 2011.

Hood, John. "Blessings of Liberty: How Business Delivers the Good." *Policy Review* 78 (July–August 1996): 14–15.

Hyman, Vicki. "The First Mass-Marketed Golf Tee." *Inside Jersey*, August 2010.

———. "A Guy Who Ran Hot and Cold." *Inside Jersey*, March 200.

———. "History Takes a Dive." *Inside Jersey*, February 2010.

———. "How New Jersey Saved Civilization by Taming Blueberries." *Inside Jersey*, July, 2010.

———. "The Joke's on Jersey." *Star-Ledger*, April 1, 2008.

———. "Play-Doh Turns 50." *Star-Ledger*, December 12, 2005.

———. "Taming the Wild Blueberry." *Inside Jersey*, June 2010.

Isaacson, Walter. "Inventing the Future." *New York Times*, April 8, 2012.

Jones, Stacy. "N.J. Frozen-Soup Maker Stirs Up a Lifesaving Recipe for Kids." *Star-Ledger*, October 2, 2011.

Kennedy, Pagan. "Who Made That Universal Product Code?" *New York Times*, January 4, 2013.

Lemelson-MIT. "James Hillier." http://web.mit.edu/invent/iow/hillier.html.

Maloney, Lawrence D. "Lone Wolf of the Sierras." *Design News*, March 6, 1995.

Materna, Peter. "Jerome Lemelson: Local Inventor and Visionary." *Nannygoats* 2, no. 1 (2004): 1, 5.

Misseck, Robert E. "Thousands Owe Lives to Cranford GI's Brainstorm." *Star-Ledger*, May 28, 2005.

Moran, Barbara. "The Preacher Who Beat Eastman Kodak." *Invention & Technology* 17, no. 2 (Fall 2001): 1–5.

Mulvihill, Geoff. "Campbell Soup Gets New Face in Its Hometown." *Courier-News*, June 11, 2010.

Noll, Michael. "A Half-Century of Info from Space." *Star-Ledger*, July 8, 2012.

Nutt, Amy Ellis. "Fame Calls on Two Titans of Telephony." *Star-Ledger*, June 19, 2008.

Perlin, John. "Photovoltaics." http://www.californiasolarcenter.org/history_pv.html.

Poundstone, William. "Unleashing the Power." *New York Times*, May 4, 2012.

Riordan, Michael, and Lillian Hoddeson. *Crystal Fire: The Birth of the Information Age*. New York: W.W. Norton, 1997.

Seckel, Alan. *Masters of Deception: Escher, Dali and the Artists of Optical Illusion*. New York: Sterling Publishing Co., 2004.

Sedieman, Tony. "Barcodes Sweep the World." http://www.barcoding.com/information/barcode_history.shtml.

Short, Simine. "Simine's U.S. Aviation Patent Database." http://invention.psychology.msstate.edu/patents/index.html.

Sive, Mary Robinson. "David T. Kenney, Unknown NJ Inventor." *Garden State Legacy* 8 (June 2010): 1–9.

Spring, Kathleen McGinn. "The Story of Color Television." *U.S.1.*, November 14, 2001.

Star-Ledger. "David Freedman, 89, A Passionate Inventor." April 28, 2010.

———. "Elihu Tchack, 84, Father of Invention." June 4, 2006.

———. "John Houghtaling, 92, Invented Vibrating Bed." June 20, 2009.

———. "Norman Gaylord, Prolific Inventor." September 25, 2007.

Thomas, Robert McG. "Jerome H. Lemelson, an Inventor, Dies at 74." *New York Times*, October 4, 1997.

Undersea Warfare. "A Disappointing Hiatus." www.navy.mil/navydata/cno/ n87/usw/issue_19/holland3.htm.

Wilk, Tom. "Ringing in the New." *New Jersey Monthly*, October 10, 2010.

Wulffson, Don L. *The Kid Who Invented the Popsicle: And Other Surprising Stories about Inventions*. New York: Puffin Books, 1997.

INDEX

ABOUT THE AUTHOR

Linda Barth has been a fan of New Jersey for a long time. As a fourth-grade teacher, she tried to focus students' attention on the positive aspects of the state: its diverse geography, agriculture, industry and famous firsts and inventions. A lifelong resident of the Garden State, Linda has served on the board of Celebrate NJ! and is active in the Delaware & Raritan Canal Watch. She has written two books about the canal, as well as two children's picture books: *Bridgetender's Boy* and *Hidden New Jersey*.

Visit us at
www.historypress.net
....

This title is also available as an e-book